MEDICAL
INTELLIGENCE
UNIT

Cardiac
Mechanotransduction

Matti Weckström, M.D., Ph.D.
Department of Physical Sciences
Division of Biophysics
and
Biocenter Oulu
University of Oulu
Oulu, Finland

Pasi Tavi, Ph.D.
Department of Physiology
Division of Biophysics
and
Biocenter Oulu
University of Oulu
Oulu, Finland

LANDES BIOSCIENCE / EUREKAH.COM
AUSTIN, TEXAS
U.S.A.

SPRINGER SCIENCE+BUSINESS MEDIA
NEW YORK, NEW YORK
U.S.A.

CARDIAC MECHANOTRANSDUCTION

Medical Intelligence Unit

Landes Bioscience / Eurekah.com
Springer Science+Business Media, LLC

Springer Science+Business Media, LLC, 233 Spring Street, New York, New York 10013, U.S.A.
http://www.springer.com

Please address all inquiries to the Publishers:
Landes Bioscience / Eurekah.com, 1002 West Avenue, Second Floor, Austin, Texas 78701, U.S.A.
Phone: 512/ 637 6050; FAX: 512/ 637 6079
http://www.eurekah.com
http://www.landesbioscience.com

ISBN 978-1-4419-2373-8   e-ISBN 978-0-387-48868-4

9 8 7 6 5 4 3 2 1

To fellow scholars

CONTENTS

EDITORS

Matti Weckström
Department of Physical Sciences
Division of Biophysics
and
Biocenter Oulu
University of Oulu
Oulu, Finland
Email: matti.weckstrom@oulu.fi
Chapter 1

Pasi Tavi
Department of Physiology
Division of Biophysics
and
Biocenter Oulu
University of Oulu
Oulu, Finland
Chapter 1

CONTRIBUTORS

Hiroshi Akazawa
Department of Cardiovascular Science
  and Medicine
Chiba University Graduate
  School of Medicine
Chiba, Japan
Chapter 6

Clive M. Baumgarten
Departments of Physiology,
  Internal Medicine (Cardiology),
  and Biomedical Engineering
Medical College of Virginia
Virginia Commonwealth University
Richmond, Virginia, U.S.A.
Email: clive.baumgarten@vcu.edu
Chapter 2

Sarah C. Calaghan
School of Biomedical Sciences
University of Leeds
Leeds, U.K.
Chapter 3

Hiroshi Hasegawa
Department of Cardiovascular Science
  and Medicine
Chiba University Graduate
  School of Medicine
Chiba, Japan
Chapter 6

Issei Komuro
Department of Cardiovascular Science
  and Medicine
Chiba University Graduate
  School of Medicine
Chiba, Japan
Chapter 6

Max J. Lab
National Heart and Lung Institute
Imperial College
London, U.K.
Email: m.lab@imperial.ac.uk
Chapter 4

Ilka Lorenzen-Schmidt
Department of Medicine
University of California, San Diego
La Jolla, California, U.S.A.
Chapter 5

Andrew D. McCulloch
Department of Medicine
University of California, San Diego
La Jolla, California, U.S.A.
Chapter 5

Jeffrey H. Omens
Department of Medicine
University of California, San Diego
La Jolla, California, U.S.A.
Email: jomens@ucsd.edu
Chapter 5

Sampsa Pikkarainen
Department of Pharmacology
  and Toxicology
and
Biocenter Oulu
University of Oulu
Oulu, Finland
Chapter 9

Hans Michael Piper
Physiologisches Institut
Giessen, Germany
Chapter 7

Heikki Ruskoaho
Department of Pharmacology
  and Toxicology
and
Biocenter Oulu
University of Oulu
Oulu, Finland
Email: heikki.ruskoaho@oulu.fi
Chapter 9

Klaus-Dieter Schlüter
Physiologisches Institut
Giessen, Germany
Chapter 7

Peter H. Sugden
NHLI Division (Cardiac Medicine)
Faculty of Medicine
Imperial College of Science, Technology
  and Medicine
London, U.K.
Email: p.sugden@imperial.ac.uk
Chapter 8

Hiroyuki Takano
Department of Cardiovascular Science
  and Medicine
Chiba University Graduate
  School of Medicine
Chiba, Japan
Chapter 6

Heikki Tokola
Department of Pharmacology
  and Toxicology
and
Biocenter Oulu
University of Oulu
Oulu, Finland
Chapter 9

Sibylle Wenzel
Physiologisches Institut
Giessen, Germany
Chapter 7

Ed White
School of Biomedical Sciences
University of Leeds
Leeds, U.K.
Email: e.white@leeds.ac.uk
Chapter 3

Yunzeng Zou
Department of Cardiovascular Science
  and Medicine
Chiba University Graduate
  School of Medicine
Chiba, Japan
Chapter 6

PREFACE

For about hundred years the investigation of heart physiology has had one central guiding principle, the "law" of Frank and Starling. This connects the return of blood into the heart and the blood pressure with cardiac contraction force. The "law" does it in a way that enables the cardiovascular system to react to perturbations without major malfunctions. This book is a compilation of reviews of prominent scientists on this subject. The difference of the original formulation of the Frank-Starling principle is that mechanotransduction is the central theme that leads the reader through the book. Since the discovery of the "law" the scope of topics related to this subject has broadened enormously, as can be seen easily by glancing at the contents of this book. Mechanotransduction in the heart has many faces that range from molecules to humans and their diseases. We editors hope that the large amount of knowledge compressed into the book's chapters forms a balanced treatment and that the text is easily approached by all who want to know what cardiac mechanotransduction is about.

Matti Weckström and Pasi Tavi
Oulu, Finland
June 16, 2006

Acknowledgments

The editors are grateful to all authors for their magnificent contributions and for their patience during compiling this book. This work was supported by the University of Oulu, the Biocenter Oulu and the Academy of Finland.

CHAPTER 1

The Mechanosensory Heart:
A Multidisciplinary Approach

Matti Weckström* and Pasi Tavi

Abstract

The cardiac muscle has an intrinsic ability to sense its filling state and react to its changes, independently of cardiac innervation that may partially serve the same functions. This ability, interesting by itself, has also a medical significance because it is associated with disturbances that may develop if a sustained loading of the myocytes will change their function. This may lead to adaptational growth of the cardiac muscle, but also to serious diseases like myocardial left ventricular hypertrophy and heart failure. In this book, and in this introductory chapter, we will focus on the nature of the sensory mechanisms of the cardiac myocytes, based on the mechanism that can be called mechanotransduction. We will look at the ability of cardiac cells to sense the filling state of the heart as a process where a mechanical stimulus is transformed into a change in the cell's functions, be it in membrane voltage, contraction force, ion balance, exocytosis or in gene expression. One possibility to do this is to divide the sensation process into limited series of more or less accurately timed events, from coding of the mechanical stimuli to both signalling via second messengers and to decoding of the information into changes in heart function, as proposed earlier.[1] As in other physiological functions, also mechanotransduction is controlled by feedback. In the heart it consists of exocytosis of vasoactive peptides and growth of the heart muscle, both tending to decrease the initial (volume) load, and of the coactivation of regulatory pathways in the nervous system. Under some circumstances the physiological regulatory loops may become maladaptive, leading to development of pathological hypertrophy and heart failure.

Introduction: Mechanosensation in the Heart

Heart is a mechanosensory organ in addition to being an efficient pump. This may have been a surprise in early 1980s to those not intimately familiar with cardiac physiology, but by now this fact is relatively well established. To appreciate the developments we first have to make clear some definitions. The concept of some cells being "mechanosensory" means that the cells have a capability of reacting to mechanical stimuli. Normally the term is applied to nerve cells or cells of sensory epithelia that transmit electrical signals triggered by mechanotransduction to other nerve cells. However, the same concept may be extended to include other kinds of transduction processes as well. When cardiac myocytes have been studied it has been found out that they are mainly sensitive to axial stretch (i.e., stretch along the axis defined by their length). They do not send electrical messages, but instead show hormonal or paracrine secretory activity with which they can communicate the message of stretch to other cells. In addition to this, they

*Corresponding Author: Matti Weckström—Department of Physical Sciences, Division of Biophysics, and Biocenter Oulu, University of Oulu, PO Box 3000, 90014 Oulun yliopisto, Oulu, Finland. Email: matti.weckstrom@oulu.fi

Cardiac Mechanotransduction, edited by Matti Weckström and Pasi Tavi.

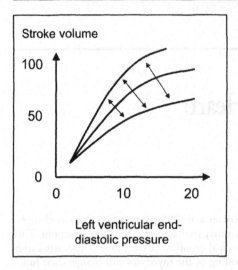

Stroke volume

100

50

0

0   10   20

Left ventricular end-
diastolic pressure

Figure 1. Frank-Starling law of the heart. The stroke volume (volume of blood ejected from the left ventricle during one contraction) in function of the volume of the left ventricle just before the contraction, i.e., in the end of the diastole (also termed LVEDP). The increase or decrease in contractile force of the ventricular musce (inotropy) change the position of the curve (arrows).

are able to react to stretch by changing their function, without any external influence like control by the nervous system.

The recent interest in heart mechanosensation began in the 1980s by the discovery of natriuretic peptides by DeBold.[2] The ability to sense stretch, and secrete, in a controlled way, hormones into the circulation was something that the heart was not generally supposed to be doing. It is, after all, a pump, keeping everything in the body alive by rapidly circulating blood and allowing exchange of gases, nutrients, metabolic products and chemical messages between different compartments of the mammalian organ systems. If cardiac myocytes can sense mechanical stimuli, they must a have a mechanism that converts—or transduces—the mechanical stimuli into some form of cellular signals. This process is clearly mechanotransduction. The basic and most obvious physiological mechanism related to mechanotransduction was described more than one hundred years ago and some mysteries of this process at cellular level have been revealed. They form a large part of the present volume. Some mysteries still remain elusive.

The mechanical sensing abilities of the heart do not—in fact—constitute a new finding as such, although the "sensing" property has not been considered in exactly this connection earlier. Mechanotransduction may be one of the basic properties of living cells, from prokaryotes, even Archaea, to advanced multicellular eukaryotes.[3-7] The history of cardiac mechano-transduction begins with the studies of two Germans, Otto Frank and Hermann Straub, and an Englishman, Ernest H. Starling, who separately defined the basic regulatory function of the heart, now known commonly as Frank-Starling law of the cardiac contraction. The works were essentially based on the isolated frog heart, a stupendously successful preparation developed in the Carl Ludwig School in Leipzig, Germany, in the late 19th century.[8] The Frank-Starling (F-S) law is a fundamental principle of cardiac behaviour, according to which the force of contraction of the cardiac muscle is proportional to the initial (precontraction) length of the muscle. This length is defined simply by the volume of the heart chamber (e.g., the left ventricle), which is again defined by the volume of blood flowing into that chamber during diastole. The force of each contraction is thus a monotonic function of prior cardiac filling (Fig. 1). When the diastolic filling of the heart is increased or decreased (like with increased or decreased venous return), the displacement of the ventricular contents (or stroke volume, the volume ejected per cycle of contraction) increases or decreases with this filling volume.[9] This basic relationship may be changed by other regulatory mechanisms, like the increase in inotropy by sympathetic stimulation acting on α- or β-receptors in the myocytes. At the time of discovery of the F-S law the cellular mechanism was a complete mystery. However, its purpose inside the big picture of the circulatory physiology was recognized

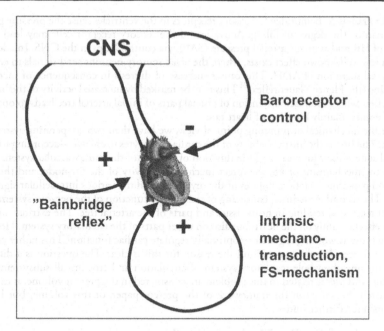

Figure 2. Systemic control loops that change heart function according to the volume that is filling the heart chambers. The Henry-Gauer reflex has been ignored.

immediately: to balance incoming volume of blood (ventricular filling) with the volume that is in fact pumped out of the heart.

The physiological and cellular origin of the F-S-mechanism has obviously been under investigation since its discovery. However, one of the reasons for the recent surge in research on cardiac mechanotransduction is the relatively new findings of coupling between cardiomyocytes' secretory activity (mainly of ANP and BNP, see reviews by refs. 10, 11 in this volume) and of cardiac hypertrophy and arrhythmias (Omens et al[12] and Lab,[13] both in this volume) to stretch of myocytes. Earliest findings on electrical changes in heart cells in response to stretch were reported by Penefsky and Hoffman[14] and Lab.[15] Thereafter considerable advance was made with the finding of two-phasic response of the myocytes to stretch,[16] and its connection to the intracellular Ca^{2+} transients[17,18] that largely define the force of contraction. Nevertheless, the multifaceted nature of cardiac mechanotransduction has made the field both difficult and interesting, as is evident from the review papers in this volume.

In addition to the FS-mechanism that affects the working muscle of the heart, the pacemaker cells in the right atria, in the SA-node, have recently been found to be in a way mechanosensitive even when isolated from the control of the nervous system.[19-22] These myocytes change their spontaneous rate of action potentials according to the stretch they are exposed to, and thereby change the heart rate. This intrinsic atrial mechanosensitivity nicely complements the FS-mechanism, allowing more room for the heart to operate. When the filling of both atria and the ventricles is increasing, the heart adapts by both pumping more forcefully (stroke volume increases) as well as increasing the rate of pumping (heart rate increases).

Before going into details of intrinsic mechanosensitivity at the cardiac myocyte level, one important aspect of the control of the heart has to be discussed shortly (Fig. 2). Other cells than cardiac myocytes but residing inside the heart tissue were also recognized to be mechanosensitive quite some time ago. Sensory innervation in the atria provides a way for the nervous control systems to sense volume of the blood (i.e., the degree of filling of the atria), and may cause, at least in experimental situations, a "reflectory" increase in heart rate (HR), named after the discoverer Francis A. Bainbridge.[23] Sensory receptors in the ventricles sense the pressure generated therein or the degree of filling. Activation of the sensory receptor cells may lead to an increase of HR and systemic arterial pressure (SAP), via control loops in the CNS. In addition to this, a less well-known effect exists, where the atrial sensory neurons send signals to control hypophyseal secretion of ADH. This causes increase of diuresis in consequence of increased atrial filling (the Henry-Gauer reflex.[24] This may be masked by increased activity of the baroreceptor reflex, where increased distension of crucial parts of initial arterial tree leads to opposite chain of events, mainly lowering of heart rate.

From the mechanical or pumping point of view, we have then two cooperating systems in the heart. The first is the intrinsic ability of the cardiac myocytes to change—according to their stretched state—their function and the function of the rest of the cardiovascular system. This includes the mechanisms of FS, the direct mechanosensitivity of the SA-node, and the hormonal and paracrine secretion abilities of the myocytes, and the various intracellular signaling systems. The second is extrinsic, consisting of the sensory function of the nervous system with nerve cell receptors, residing in heart tissue and parts of the arterial tree. The extrinsic system exerts its effects mainly on the heart, but also on other parts of the circulatory system. How are these two systems working together to optimally regulate cardiac function? This rather crucial question has been largely ignored, and the reason for this is clear. The question is a difficult one, comprising the whole regulatory system of circulation and crossing all subsystems. Although bits and pieces related to this problem are considered in the present volume, it cannot, unfortunately, be solved in the framework of the present paper, or this volume, but has to remain open for further studies.

Cornucopia of Cardiac Mechanotransduction

In the present volume we are concentrating on different aspects of cardiac mechanosensitivity. Being a complex issue, many differents disciplines are involved, from biophysics and biochemistry to physiology and clinical sciences. Not surprisingly, large interest has been placed on the initial mechanisms of intrinsic mechanotransduction in cardiac myocytes. Simple logic, starting from calcium dependence of myocyte contraction and its initiation by an electrical event, the action potential, has lead to a search for mehanosensitive ion channels. In fact numerous channels have been found, exhibiting mechanosensitivity (Baumgarten,[25] this volume). Unfortunately, the issue is fairly complex and no ubiquitous source for the stretch-sensitivity of the myocytes have been found so far. Instead, mechanosensitivity in toto seems to consist of several overlapping and parallel mechanisms, that may, or may not, have a common initial source. These mechanisms entail changes in the contractile apparatus of the heart (Calaghan and White,[26] this volume), and lead to a concept of mechano-electric feedback, a process that complements the sequence of events from action potential to contraction, the electro-mechanical coupling (Lab,[15] this volume). Whether or not initiated by ion channel activations, mechanotransduction involves a complex modulation of intracellular enzymatic signaling pathways, including those stimulated by external, hormonal stimulation (Hasegawa et al,[27] Schlüter et al,[28] Sugden,[29] this volume). The hypothesis can be advanced, though, that the initial events in mechanotransduction are linked to ion channel activation. Thus it is of importance, that the mammalian cation selective mechano-sensitive channel has recently been identified.[30] It is a member of the TRP-channel family, initially implicated in phototransduction in Drosophila photoreceptors, and seems to be related also to the C. elegans proteins that are involved in mechanotransduction. Whether this channel forms the basis of cardiac mechanotransduction, remains to be seen.

Considering the many faces of cardiac mechanotransduction (Fig. 3) it is not surprising that it is also implied in long term adaptive or maladaptive changes of heart function.[1] This is manifest in the mechanical adaptations[12] as well as in many targets of the involved signalling pathways.[27-29] Finally, the circuit is closed by taking into account the signals that the heart sends either to itself (paracrinic control) or to other cells by secreting several hormones in

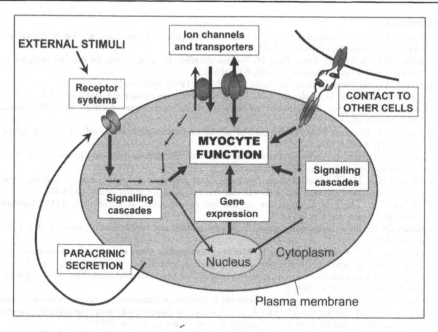

Figure 3. A diagram showing the multiple mechanisms contributing to the cardiac myocyte's response to mechanical stimulation; for more details see ref. 1.

response to mechanical stimulation.[11] This leads to at least partial relaxation of those processes that cause stretch, of mechanical loading, of the heart, and comprise a negative feedback from mechanotransduction to its causes.

One great problem that encounters an investigator willing to unravel the mysteries of cardiac physiology is the multiplicity of processes involved. Consider the ion channels: almost all types of channels found are also expressed in cardiac myocytes. Consider the signaling pathways: most hormonal and paracrine signals known to have receptors in the heart and concomitantly also the complex network of signaling pathways are present.[1,31] Consider the adaptational ability: the heart can function as dilated "athlete's heart",[32,33] fast and strong but using large amounts or energy, or dilated but slow and weak,[34,35] but using lesser amounts of energy. The multi-faceted nature of the physiology of cardiac myocytes is reflected necessarily in their mechanotransduction. Being a mechanical pump, able to rapidly adapt to circumstances, mechanical loading of the myocytes rapidly changes their function. This may lead to long lasting effects as well, via changes in gene expression of crucial proteins, including those in the contractile apparatus and in the signaling pathways. The present volume contains reviews of much of the recent work in cardiac mechanotransduction but it most certainly is not the final word. A lot remains to be found. Most importantly in our view, the initial events in transducing the mechanical load, or stretch, into changes in physiology of the cells needs to be known. Following this findings it would be easier to put all the other effects into perspective, possibly via regulation of intracellular calcium balance,[36] calcium-activated signaling pathways,[1,31] possibly consequent or parallel changes in other signaling mechanisms,[27-29] and finally the changing patterns in regulation of gene expression. The same importance has to be applied to the holistic aspects of cellular mechanotransduction: how are all the regulatory properties integrated into a meaningful, adaptive system. This knowledge (or understanding) would be very useful in applications, i.e., in developing prevention, diagnosis and treatment of cardiac load-related diseases.

References

1. Tavi P, Laine M, Weckström M et al. Cardiac mechanotransduction: From sensing to disease and treatment. Trends Pharmacol Sci 2001; 22:254-60.
2. Baines AD, DeBold AJ, Sonnenberg H. Natriuretic effect of atrial extract on isolated perfused rat kidney. Can J Physiol Pharmacol 1983; 61:1462-6.
3. French AS. Mechanotransduction. Annu Rev Physiol 1992; 54:135-152.
4. Hamill O, McBride D. Molecular clues to mechanosensitivity. Biophys J 1993; 65:17-8.
5. Ingber DE. Tensegrity: The architectural basis of cellular mechanotransduction. Annu Rev Physiol 1997; 59:575-99.
6. Hamill OP, Martinac B. Molecular basis of mechanotransduction in living cells. Physiological Reviews 2001; 81:685-740.
7. Kloda A, Martinac B. Common evolutionary origins of mechanosensitive ion channels in Archaea, Bacteria and cell-walled Eukarya. Archaea 2002; 1:35-44.
8. Zimmer HG. Modifications of the isolated frog heart preparation in Carl Ludwig's Leipzig Physiological Institute: Relevance for cardiovascular research. Can J Cardiol 2000; 16:61-69.
9. Starling EH. The LINACRE lecture on the law of the heart, at Cambridge, 1915. London: Longmans, Green and Co., 1918.
10. Ruskoaho H. Atrial natriuretic peptide: Synthesis, release, and metabolism. Pharmacol Rev 1992; 44:479-602.
11. Pikkarainen S, Tokola H, Ruskoaho H. Mechanotransduction of the endocrine heart paracrine and intracellular regulation of B-type natriuretic peptide synthesis. In: Weckström M, Tavi P, eds. Cardiac Mechanotransduction. Georgetown: Landes Bioscience/Eurekah.com; New York: Springer Science+Business Media, 2006:9.
12. Omens JH, McCulloch AD, Lorenzen-Schmidt I. Mechanotransduction in cardiac remodeling and heart failure. In: Weckström M, Tavi P, eds. Cardiac mechanotransduction. Georgetown: Landes Bioscience/Eurekah.com; New York: Springer Science+Business Media, 2006:5.
13. Lab MJ. Lab mechanoelectric transduction/feedback: physiology and pathophysiology. In: Weckström M, Tavi P, eds. Cardiac Mechanotransduction. Georgetown: Landes Bioscience/Eurekah.com; New York: Springer Science+Business Media, 2006:4.
14. Penefsky ZJ, Hoffman BF. Effect of stretch on mechanical and electrical properties of cardiac muscle. Am J Physiol 1963; 204:433-438.
15. Lab MJ. Mechanically dependent changes in action potentials recorded from the intact frog ventricle. Circ Res 1978; 42:519-528.
16. Parmley WW, Chuck L. Length-dependent changes in myocardial contractile state. Am J Physiol 1973; 224:1195-1199.
17. Allen DG, Kurihara S. The effects of muscle length on intracellular calcium transients in mammalian cardiac muscle. J Physiol 1982; 327:79-94.
18. Allen DG, Nichols CG, Smith GL. The effects of changes in muscle length during diastole on the calcium transient in ferret ventricular muscle. J Physiol 1988; 406:359-370.
19. Hagiwara N, Masuda H, Shoda M et al. Stretch-activated anion currents of rabbit cardiac myocytes. J Physiol 1992; 456:285-302.
20. Horner SM, Murphy CF, Coen B et al. Contribution to heart rate variability by mechanoelectric feedback. Stretch of the sinoatrial node reduces heart rate variability. Circulation 1996; 94:1762-7.
21. Arai A, Kodama I, Toyama J. Roles of Cl⁻ channels and Ca^{2+} mobilization in stretch-induced increase of SA node pacemaker activity. Am J Physiol 1996; 270:H1726-35.
22. Cooper PJ, Lei M, Cheng LX et al. Selected contribution: Axial stretch increases spontaneous pacemaker activity in rabbit isolated sinoatrial node cells. J Appl Physiol 2000; 89:2099-104.
23. Hakumäki MO. Seventy years of the Bainbridge reflex. Acta Physiol Scand 1987; 130:177-85.
24. Gauer OH, Henry JP, Behn C. The regulation of extracellular fluid volume. Annu Rev Physiol 1970; 32:547-595.
25. Baumgarten CM. Origin of mechanotransduction: Stretch-activated ion channels. In: Weckström M, Tavi P, eds. Cardiac Mechanotransduction. Georgetown: Landes Bioscience / Eurekah.com; New York: Springer Science+Business Media, 2006:2.
26. Calaghan SC, Ed White E. The role of the sarcomere and cytoskeleton in cardiac mechanotransduction. In: Weckström M, Tavi P, eds. Cardiac mechanotransduction. Georgetown: Landes Bioscience / Eurekah.com; New York: Springer Science+Business Media, 2006:3.
27. Hasegawa H, Takano H, Zou Y et al. Second messenger systems involved in heart mechano-transduction. In: Weckström M, Tavi P, eds. Cardiac mechanotransduction. Georgetown: Landes Bioscience/Eurekah.com; New York: Springer Science+Business Media, 2006:6.

28. Schlüter K-D, Piper HM, Wenzel S. The role of adrenoceptors in mechanotransduction. In: Weckström M, Tavi P, eds. Cardiac mechanotransduction. Georgetown: Landes Bioscience/Eurekah.com; New York: Springer Science+Business Media, 2006:7.
29. Sugden PH. Intracellular signaling through protein kinases in cardiac mechanotransduction. In: Weckström M, Tavi P, eds. Cardiac mechanotransduction. Georgetown: Landes Bioscience/Eurekah.com; New York: Springer Science+Business Media, 2006:8.
30. Maroto R, Raso A, Wood TG et al. TRPC1 forms the stretch-activated cation channel in vertebrate cells. Nat Cell Biol 2005; 7:105-7.
31. Weng G, Bhalla US, Iyengar R. Complexity in biological signaling systems. Science 1999; 284:92-6.
32. Fagard R. Athlete's heart. Heart 2003; 89:1455-61.
33. Cantwell JD. The athlete's heart syndrome. Int J Cardiol 1987; 17:1-6.
34. Morgan HE, Baker KM. Cardiac hypertrophy. Mechanical, neural, and endocrine dependence. Circulation 1991; 83:13-25.
35. Wikman-Coffelt J, Parmley WW, Mason DT. The cardiac hypertrophy process. Analyses of factors determining pathological vs. physiological development. Circ Res 1979; 45:697-707.
36. Bers DM, Edward Perez-Reyes. Ca channels in cardiac myocytes: Structure and function in Ca influx and intracellular Ca release. Cardiovasc Res 1999; 42:339-360.

CHAPTER 2

Origin of Mechanotransduction:
Stretch-Activated Ion Channels

Clive M. Baumgarten*

Abstract

S tretch-activated ion channels (SAC) serve as cardiac mechanotransducers. Mechanical stretch of intact tissue, isolated myocytes, or membrane patches rapidly elicits the opening of poorly selective cation, K^+, and Cl^- SAC. Several voltage- and ligand-gated channels also are mechanosensitive. SAC alter cardiac electrical activity and, with prolonged stretch, cause an intracellular accumulation of Ca^{2+} and Na^+ that can serve to trigger multiple signaling cascades and ultimately may contribute to remodeling of the heart in response to hemodynamic stress. This chapter reviews the transmission of mechanical forces, the biophysical characteristics of cardiac SAC, and how SAC activity may be coupled to signaling cascades and thereby initiates the complex response of the heart to stretch.

Introduction

Mechanical forces impinging on the heart play multiple roles in controlling cardiac function.[1,2] It is well-known that physiological mechanical stimuli, such as ventricular end diastole volumes, regulate the force of cardiac contraction and that atrial stretch regulates the secretion and production of atrial and brain natriuretic peptides. Transient mechanical stimuli alter cardiac electrical activity and rhythm, whereas sustained stimulation, for example, pressure or volume overload or wall motion abnormalities, activates a complex set of signaling cascades and gene programs leading to profound remodeling of the heart.

Cardiac responses to mechanical stimuli clearly are diverse. Yet, each process must begin with mechanotransduction, the conversion of mechanical force to electrical or biochemical signals that ultimately control cellular function. This chapter focuses on one instrument of the transduction process, stretch-activated channels (SAC)

SAC typically respond to mechanical perturbations within tens of milliseconds. Their activity is the initial response of the heart to stretch and can serve to trigger subsequent events. An overview of how SAC translate mechanical stimuli is presented in Figure 1. SAC alter cardiac function by two direct mechanisms. First, SAC directly affect electrical activity by virtue of the current they pass. This leads to alterations in action potential duration and shape, depolarization of the resting potential, and in some instances, stretch-induced automaticity.[3] In turn, the altered action potential waveform and resting potential influence the behavior of virtually all voltage-dependent channels. The effects of stretch on cardiac electrical activity are called mechanoelectrical feedback.

*Clive M. Baumgarten—Departments of Physiology, Internal Medicine (Cardiology), and Biomedical Engineering, Medical College of Virginia, Virginia Commonwealth University, 1101 E. Marshall St., Richmond, Virginia 23298-0551, U.S.A. Email: clive.baumgarten@vcu.edu

Cardiac Mechanotransduction, edited by Matti Weckström and Pasi Tavi.
©2007 Landes Bioscience and Springer Science+Business Media.

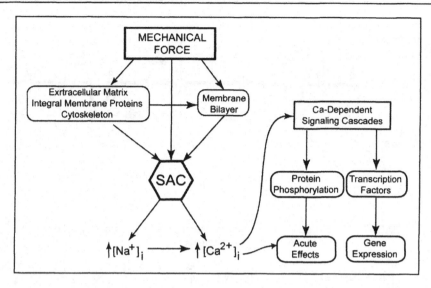

Figure 1. Overview of mechanotransduction by stretch-activated channels (SAC). Mechanical forces are transmitted to SAC by extracellular matrix, integral membrane proteins, cytoskeleton, and perhaps, the membrane bilayer. The current carried by SAC has direct effects on cardiac electrophysiology (not shown) and leads to accumulation of $[Na^+]_i$ and $[Ca^{2+}]_i$. Ca^{2+} activates multiple signaling cascades that have acute effects via protein phosphorylation and long-term effects via transcription factors and altered gene expression.

Activation of SAC also directly leads to an ion influx that alters intracellular Ca^{2+} ($[Ca^{2+}]_i$) and Na^+ ($[Na^+]_i$) concentrations,[4] and elevation of $[Na^+]_i$ favors additional cellular Ca^{2+} accumulation via Na^+-Ca^{2+} exchange. Ca^{2+} acts as a signaling molecule to switch on Ca^{2+}-dependent signaling cascades including PKC, PyK, calmodulin, CaM kinases, and phospholipases that modulate cardiac function by protein phosphorylation and ultimately by stimulating transcription factors and protein synthesis.[5,6]

Terminology

The term stretch-activated channel is adopted because it is widely used in the cardiac literature. More broadly, channels responding to mechanical forces should be termed mechanosensitive channels (MSC), which recognizes that channels can be activated or inactivated by stretch. Channels that respond to stretch, curvature, or deformation of the membrane are classified as mechanosensitive, even though the details of how a particular stimulus leads to altered channel function is rarely known. Osmotic or hydrostatic swelling or shrinkage of cells also regulate certain channels. Whereas these maneuvers stretch the membrane, it is unclear whether mechanical forces or another aspect of a volume change, such as dilution (or concentration) of cytoplasmic ions and macromolecules, is the stimulus. The term volume-sensitive channel will be used to differentiate between stimuli and, perhaps, underlying mechanisms. A distinction between mechano- and volume-sensitivity often has not been made in the literature, and this is a point of confusion. Cardiac volume-sensitive channels are reviewed elsewhere[7-9] and will not be discussed in detail here. Finally, a number of voltage- and ligand-gated channels also are modulated by mechanical forces. This should not be surprising because most channels can be regulated by multiple mechanisms.

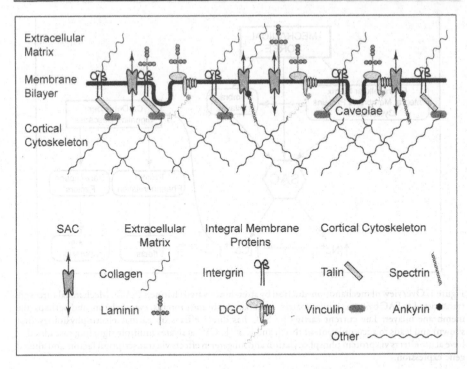

Figure 2. Diagram of membrane bilayer and components of extracellular matrix (ECM) and cortical cytoskeleton that transmit mechanical forces in myocytes. See figure for key. Hypothetical connections between SAC and both the ECM and cytoskeleton also are shown. SAC may or may not bind ECM proteins (laminin and collagen) directly, may or may not connect to the cytoskeleton via ankyrin and spectrin, and may or may not be associated with caveolae, integrins, or the dystrophin-dystroglycan complex (DGC). DGC includes the integral membrane proteins β-dystroglycan and the sarcoglycans, the extracellular protein α-dystroglycan, and dystrophin, which links the DGC to cytoskeletal actin. For review of cytoskeleton, see references 10, 11 and 14.

How Are Mechanical Forces Transmitted in the Heart?

Forces that gate channels ultimately must be transmitted by the membrane bilayer, cytoskeleton, extracellular matrix, or a combination of these elements, as illustrated in Figure 2. Cardiac myocytes possess specialized structures for transmitting force from one myocyte to another and from the extracellular matrix (ECM) to the membrane and cytoskeleton.[10,11] These structures also serve to organize and initiate signaling cascades.[12-14] The membrane bilayer is supported by a tightly associated, submembrane (cortical) cytoskeleton that remains in place even after excision of a membrane patch. Collagenase-based myocyte isolation methods strip off components of the extracellular matrix, but laminin remains in a raster-like array on the sarcolemma and remnants of fibronectin are present.[15] ECM also is present on cultured cardiac myocytes.[16] As of yet, it is impossible to determine how each of these components influences the response of cardiac SAC to stretch. In contrast, cloned prokaryotic mechanosensitive channels can be functionally reconstituted in a protein-free lipid bilayer and are controlled by bilayer curvature, bilayer thickness, and the area occupied by the protein.[17,18] This does not establish that force transmission through the bilayer is the only mechanism of activation in situ, however. *Unc-105*, a mechanosensitive *C. elegans* channel, binds to collagen IV and is thought to be activated by forces transmitted via ECM.[19]

It is reasonable to think that the cytoskeleton plays a role SAC activation. A number of ion channels and transporters (e.g., Na^+ channels, Na^+-K^+ pump, Na^+-Ca^{2+} exchanger, Cl^--HCO_3^- exchanger) and probably SAC are coupled to the cortical cytoskeleton by ankyrins, a family of peripheral membrane proteins.[11,20] All three ankyrin genes (*Ank1-3*) are expressed in human heart and give rise to a multitude of protein isoforms by alternative splicing. Furthermore, a number of ion channels contain ankyrin-like binding domains.

At the level of the myocyte, force is transmitted at costameres, membrane specializations at the each Z-line and M-line that provide a physical link between the contractile apparatus and the cytoskeleton, membrane, and ECM.[14,21] Costameres contain integrins, heterodimeric transmembrane proteins that are receptors for extracellular matrix proteins and connect to the cytoskeleton via talin, vinculin, and α-actinin. Integrins act as mechanotransducers[22] and also act as signaling molecules, especially through focal adhesion kinase (FAK).[23,24] A second system for linking ECM, membrane, and cytoskeleton is the dystrophin-dystroglycan complex (DGC).[25] Dystroglycan is a transmembrane receptor for laminin and its cytoplasmic tail binds to costameric actin. Dystroglycan and the DGC form a strong mechanical link, and dystroglycans are important for organizing other components of the DGC, as well as for assembly of vinculin and spectrin into the costameres. Loss of DGC components and resulting sarcolemmal fragility underlies muscular dystrophies.[26] Furthermore, gating of skeletal muscle SAC is altered in dystrophin-deficient *mdx* mice.[27,28]

In the intact heart, myocytes are coupled to each other by N-cadherins, transmembrane adhesion molecules that dimerize between cells. N-cadherins are found at the fascia adherens, which joins the actin cytoskeleton to the membrane, and at desmosomes, which attach intermediate filaments to the intercalated disc.[11] This physical attachment between cells is disrupted during myocyte isolation, and the potential role of N-cadherins in cardiac mechanosensitivity has not been studied.

Characteristics of Mechanosensitive Currents in Single Channel Recordings

It has long been thought that specialized mechanosensory cells were likely to possess mechanosensitive ion channels as detectors, but the need for SAC in most other types of cells was not obvious. Wide-spread interest was kindled by the discovery that a poorly-selective cation channel is activated in skeletal muscle by stretching the sarcolemma when suction is applied to a patch pipette,[29,30] as diagramed in Figure 3. With physiological Na^+ and K^+, the I-V relationship of the cation SAC in skeletal muscle is nearly linear, with a unitary conductance of 35 pS, and reverses at -30 mV; P_K:P_{Na} is 4 based on reversal potential (E_{rev}) and 2 based on conductance.[30] As negative pressure is increased from 0 to -50 mm Hg, thereby curving the sarcolemma outward and increasing tension on the membrane bilayer and adherent cortical cytoskeleton, cation SAC open probability increases ~4 orders of magnitude. Kinetic analysis of single channel open and closed times is consistent with a model with 3 closed and 1 open state. Stretch affects only a single rate constant, slowing the $C_1 \rightarrow C_2$ transition moving away from the open state, which leads to increased open probability.[29] Based on Eyring rate theory and the behavior of elastic materials, the high sensitivity of open probability to membrane tension suggests that the SAC gathers energy from a region 500 to >2000 Å in diameter.[29,31] This is far too large an area to be occupied by the SAC itself and leads to the conclusion that the submembrane cytoskeleton must participate in mechanotransduction. Membrane lipid flows into patches held under negative pressure, and patch area and capacitance increase in parallel over time, and therefore, tension in the bilayer is not sustained.[32]

Several types of cardiac SAC are found in cell-attached and excised patches from atrial and ventricular myocytes using the same pipette suction method. These include K^+-selective, poorly selective cation, and anion SAC, and mechano-gated ATP-sensitive K^+ channels (K_{ATP}), muscarinic K^+ channels (K_{ACh}), and Ca^{2+}-activated K^+ channels (K_{Ca}) also have been identified. Examples of single channel records from cardiac SAC are shown in Figure 3A-C. Kinetic analysis

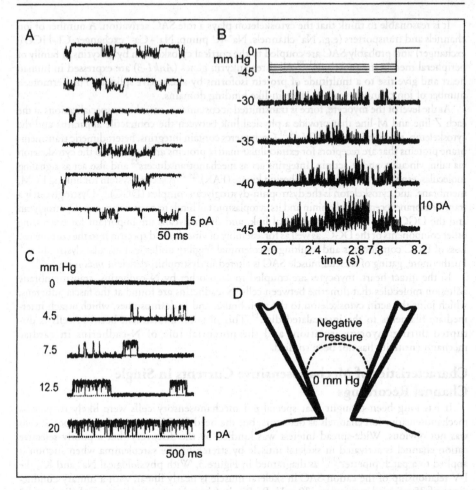

Figure 3. Unitary currents from cardiac SAC. A) Cation SAC recorded from on-cell patch on chick myocyte. (Modified from ref. 57 with permission from Lippincott Williams & Wilkins, http://www.lww.com.) B) K⁺ SAC from on-cell patch on rat atrial myocyte rapidly respond to pressure step (rise time 2-4 ms). Open probability increases with increased negative pressure. Dashed lines indicate onset and termination of pressure pulse. (Modified from ref. 44.) C) Cl⁻ SAC recorded from outside-out patch on human atrial myocyte. Increasing positive pipette pressure increases open probability in the outside-out configuration. Dotted line indicates closed state. (Modified from ref. 34.) D) Schematic diagram of effect of negative pipette pressure on cell membrane in on-cell patch. Suction causes membrane to take on a hemispherical shape. The increase in apparent membrane area is due to membrane unfolding and flow of additional lipids into the patch.[32]

of cardiac SAC reveals a more complex picture than in skeletal muscle; depending on the channel, there can be multiple open and closed states, and the open times, closed times, or both are modulated by stretch.[33-35]

As fruitful as the pipette suction method has been, it remains uncertain whether the detailed properties of SAC in patches accurately reflect their properties in intact cells or tissues. Patch formation disrupts the connections between the cortical cytoskeleton and its more central members, which may modify the forces sensed by channels and their regulation by signaling molecules. Additionally, variable patch geometry complicates interpretations because it is likely that tension rather than applied pressure controls SAC activity. In addition, the effect of patch formation on caveolae is unknown. These specialized membrane lipid rafts are omega-shaped in cross-section, are estimated to comprise 20-30% of sarcolemmal area in atria,[36] and are an important site of signaling integration.[37] The shape of the caveolae and its association with cytoskeletal elements and integral membrane proteins suggest that caveolae might be responsive to imposed forces. Caveolae are osmotically responsive[38] and appear to be incorporated into the membrane by increasing ventricular volume,[39] although earlier quantitative studies in atria failed to detect stretch-induced unfolding of caveolae over a physiologic range of sarcomere lengths.[36]

K⁺ SAC

Cardiac K⁺ SAC were first reported in snail ventricular[40,41] and also are present in embryonic chick ventricle[33,35,42] and adult rat atrium[43-45] and ventricle.[46] Unitary conductance ranges from 25 to 200 pS, indicating multiple K⁺ SAC coexist, even in the same preparation.[33,35] In chick heart, for example, the I-V relationships for 100 and 200 pS K⁺ SAC in cell-attached patches are linear with either K⁺ or Na⁺ in the pipette, i.e., with either a symmetrical or physiological ionic gradient.[33] On the other hand, the dominant K⁺ SAC in rat heart has a conductance of ~100 pS in symmetrical K⁺, and the I-V relationship is outwardly rectifying with both a physiological and a symmetrical gradient.[43-45]

There is variability in the negative pressure required to maximally activate K⁺ SAC both between investigators and within a single study.[43,44,46] This has been explained by arguing that tension in the bilayer or cortical cytoskeleton rather than the measured transmembrane pressure activates SAC.[32] Variability then arises from differences in patch radius, which determines the relationship between pressure and tension by the Law of Laplace, and perhaps from the condition of the attached cytoskeleton.

The kinetics of the stretch response were studied by Niu and Sachs[44] using a pressure clamp capable of changing pipette pressure in 2-4 ms. K⁺ SAC open with a latency of ~200 ms, and open probability undergoes adaptation with a time constant of ~1 s. Channels close within 50 ms on the release of negative pipette pressure, and spontaneous openings are suppressed for ~1 s. These data are consistent with the notion that force is coupled to the channel by a viscoelastic element, perhaps representing rearrangement of cytoskeletal elements over time, and that adaptation during continued negative pressure reflects a decrease in the force sensed by the channel. Alternatively, both the adaptation during suction and suppression of spontaneous events upon release can be modeled by including an inactivated state in the kinetic model of the channel. Transitions from the open to the inactivated state would reduce openings in the presences of a constant stimulus, and upon termination of the stimulus, fewer channels would initially return to the closed state and so basal activity would be suppressed.

Recent studies identified molecular candidates for K⁺ SAC: TREK-1, TREK-2 and TRAAK. These channels are members of the tandem two-pore domain (2P) K⁺ channel family (KCNK2, KCNK10, and KCNK4, respectively) that also includes TWIK, THIK and TALK (for review, see refs. 47-49). Unfortunately, none of the known mechanosensitive 2P K⁺ channel isoforms is detected in human heart,[47,50] but TREK-1 is expressed in mouse heart[51] and in rat atrial and ventricular myocytes.[45,46] Heterologously expressed TREK-1 (KCNK2) opens in flickering bursts with a unitary conductance of ~100 pS in symmetrical K⁺, is reversibly opened by polyunsaturated fatty acids including arachidonate, and intracellular acidosis both activates the channel directly and sensitizes it to stretch.[52] These properties mirror those of rat K⁺ SAC[43-46] and support the argument that endogenous K⁺ SAC may be TREK-1.

Cation SAC

Pipette suction also activates multiple poorly selective cation SAC. These channels have been studied in neonatal and cultured rat ventricle,[53,54] neonatal and adult rat atria,[55,56] guinea pig ventricle,[57] embryonic chick ventricle,[33,42,57] and endocardial endothelial cells in porcine atria.[58] As shown for K⁺ SAC, multiple cation SAC are present in the same preparation.[33] Unitary conductances cluster around 21 to 25 pS, but 50 and 120 pS channels are also found. Based on electrophysiological data, cation SAC poorly distinguish between monovalent cations including Na^+, K^+, Cs^+, Rb^+, and Li^+,[33,55,56,58] and reported $P_K:P_{Na}$ ratios vary from 0.7 to ~5. Divalent cations including Ca^{2+}, Ba^{2+}, and Mg^{2+} also are permeant.[33,55,57] Conductance for Ca^{2+} was 80% of that of K^+ with equimolar replacement[55] and 40% with equal equivalents (i.e., 140 mM K^+ vs. 70 mM Ca^{2+}).[58] Therefore, Na^+ is the dominant charge carrier of inward cation SAC current with a physiological extracellular ionic milieu.

Cation SAC are fully activated by 20-30 mm Hg of suction,[33] but another study reported full activation at 4 mm Hg.[55] This difference may reflect methodological issues that affected tension.

Members of the TRP family of Ca^{2+}-permeant cation channels are candidates for the cation SAC.[59,60] In particular, TRPV4, a mammalian vanillinoid receptor-like TRP, is osmosensitive and is a homolog of *C. elegans* (OSM-9) and *Drosophilia* (NOMPC) channels that are osmo- and mechanosensitive. However, TRPV4 is not expressed in heart.[61] *

Pharmacology of K^+ and Cation SAC

Yang and Sachs[62] identified the lanthanide Gd^{3+} as a blocker of SAC in *Xenopus* oocytes, and this lanthanide also blocks most cardiac cation and K^+ SAC at 10-30 µM.[33,63] Gd^{3+} inhibits stretch-activated TRAAK,[64] lysolipid-activated TREK-1,[65] but not the background activity of TREK-1 even at 100 µM.[51] Gd^{3+} has been widely adopted as a tool for identifying the physiological effects of SAC, but several cautions are in order. As noted in the original report, Gd^{3+} blocks multiple transport processes. In heart, nonspecific effects of Gd^{3+} include block of I_{Na},[66] I_{Ca-L},[54,67] I_{Kr},[68] and Na^+-Ca^{2+} exchange.[69] In addition, Gd^{3+} avidly binds to EGTA and physiologic anions such as phosphate and bicarbonate.[70] These nonspecific and anion-binding properties complicate interpretation of experiments with Gd^{3+}. Examples of Gd^{3+}-insensitive cation[55,56] and K^+ SAC[43] also have been reported. In two of these studies,[43,55] EGTA apparently was included with Gd^{3+}, and therefore, the free-Gd^{3+} concentration is likely to have been too low to be effective.[70]

Streptomycin and other cationic aminoglycoside antibiotics, such as neomycin and gentamicin, and amiloride and its derivatives block SAC in a variety of tissues (for review, see ref. 71). Streptomycin inhibits a poorly selective cardiac cation current and Ca^{2+} influx in response to stretch,[72,73] but block of SAC by streptomycin apparently has not been studied at the single channel level in heart. Like Gd^{3+}, streptomycin is nonspecific and blocks other ion channels including I_{Ca-L}, I_{Kr}, and I_{Ks}.[74]

Sachs and coworks discovered that the venom of a Chilian tarantula commonly called *Grammastola spatulata* (but reclassified as *Phrixotrichus spatulatus*[75]) blocks both a 21 pS cation SAC and a 90 pS K^+ SAC in chick ventricular myocytes,[42] as well as SAC and Ca^{2+} influx into GH_3 cells upon hyposmotic swelling.[76] An active peptide, GsMTx4, was purified by screening against an inwardly-rectifying cation SAC in patches on astrocytes, and GsMTx4 fully blocked a persistently-activated, cell volume-regulated, inwardly-rectifying cation SAC in ventricular myocytes from a rabbit aortic regurgitation heart failure model[77] and stretch-induced atrial fibrillation in isolated rabbit hearts.[78] GsMTx4 appears to be potent (blocks at 170-400 nM) and highly selective for SAC and is likely to be a useful tool for studying SAC-dependent processes in the future.

Cl⁻ SAC

Volume-sensitive Cl^- currents are well known in the heart,[9] but the contribution of anions to stretch-activated currents often has been doubted. Nevertheless, Sato and Koumi[34] described an 8.6 pS Cl^- SAC in excised patches from human atrial myocytes. Activation occurs with positive pipette pressure in the outside-out and negative pressure in the inside-out configurations, and open probability increases from 0.03 at 4.5 mm Hg to 0.94 at 20 mm Hg. Unitary currents are blocked by 9-anthracene carboxylic acid (9AC) and DIDS with the same potency as the

---

* See *Notes Added in Proof.*

volume-sensitive whole-cell Cl⁻ currents in these cells. An outwardly-rectifying mechanosensitive Cl⁻ channel with a unitary conductance of 5-7 pS is described in *C. elegans* embryonic cells.[79] This channel also fully activates with -20 mm Hg of pipette pressure and exhibits I⁻ > Cl⁻ permeability, as do volume-sensitive Cl⁻ channels in heart.[9] In contrast to volume-sensitive Cl⁻ currents, *C. elegans* Cl⁻ SAC undergo delayed rectification at positive voltages rather than inactivation.

Mechanosensitive Ligand-Gated Channels

Van Wagoner[80] demonstrated that K_{ATP} channels in neonatal rat atrial myocytes are stimulated by negative pipette pressure in on-cell and excised patches. Open probability increase from ~0.01 to 0.7 at -27 mm Hg. This 52-pS channel was identified by its block by ATP and tolbutamide, a sulfonyl urea K_{ATP} inhibitor, and the potentiation of mechanosensitivity by pinacidil, a K_{ATP} channel opener, at concentrations too low to enhance basal channel activity. Stretch also reactivates K_{ATP} channels that have undergone rundown but does not overcome block by 5 mM ATP. Interestingly, ischemia potentiates mechanosensitivity of K_{ATP} channels.[81]

Muscarinic K_{ACh} channels in neonatal rat atria are modulated by membrane stretch, but only in the presence of ACh.[82] Negative pressures up to -80 mm Hg with 10 μM ACh in the pipette increase open probability in a graded fashion by ~2-fold, a more modest stimulation than of other SAC. Although K_{ACh} is regulated by G protein, G_K, several lines of evidence argue that the effects of stretch are independent of the ACh receptor and G_K. Stretch of inside-out patches enhances channel activity after maximal activation of G_K with GTPγS but does not alter sensitivity to GTP. In addition, stretch modulates K_{ACh} channels activated by trypsin via a G protein-independent pathway.

Kawakubo et al[35] recently reported on a Gd^{3+}-sensitive, 200 pS K⁺ SAC in embryonic chick ventricle that is similar to the 200 pS channel described previously.[33] These authors concluded the channel is a Ca^{2+}-activated K⁺ channel activated by ATP ($K_{Ca,ATP}$). As typical for Ca^{2+}-activated K⁺ channels, charybdotoxin and TEA block $K_{Ca,ATP}$.

Characteristics of Mechanosensitive Currents in Whole-Cell Recordings

Applying a controlled, reproducible, and well-defined mechanical force to an isolated myocyte is technically challenging, and whether the methods adopted by investigators to stretch myocytes are physiologically relevant remains an open question. Given the geometry of the heart, the rotation of the alignment of myocytes going from endocardium to epicardium, and the varied mechanical disturbances introduced by cardiac pathologies, the force vectors impinging on myocytes in different situations are complex and not simple to reproduce experimentally on isolated myocytes. Moreover, because experimental tools are attached to myocytes by nonphysiological means, the transmission of applied forces and the response of the cell may or may not duplicate what transpires in situ.

Methods of Stretching Myocytes

Multiple techniques have been developed to apply force to cardiac myocytes (for references, see refs. 83,84), but only a subset of these have used to study mechanosensitive whole-cell currents, as illustrated in Figure 4. The effects of axial stretch of myocytes on whole-cell currents were first investigated by Sasaki et al[85] by fixing one end of the cell to a polylysine-coated glass coverslip and the other end to a glass microtool (20 μm diam. ball) or suction pipette. By moving the tool or suction pipette with a micromanipulator, a controlled and measurable stretch was obtained. Axial stretch involving most of the myocyte also can be generated by stretching a cell between two patch[56] or between functionalized concentric double pipettes[86] or carbon filaments[73,87,88] that adhere to the membrane because of their surface charge. Localized axial stretch between a microtool and a nearby patch pipette also has been examined.[89,90] Typically, the entire myocyte or a selected region is stretched by 5-20%, as measured from the distance between the tools or from analysis of sarcomere length. Although these methods are designed to produce a uniform stretch, nonaxial force vectors are expected near the attachment points as forces are transmitted outward from the attachment point and towards the opposite

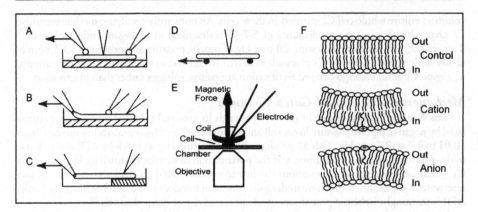

Figure 4. Methods for stretching isolated myocyte whole recording whole-cell currents. Sasaki et al[85] stretched myocytes resting on agar (A) between two rounded (20 μm diam.) glass tools or (B) between a glass tool and a suction pipette (diam. 8-10 μm, -20 to -40 mm Hg), or (C) between a polylysine-coated glass pedestal and a suction pipette. (A-C; modified from ref. 85.) D) Le Guennec et al[87] developed a method for stretching myocytes between stiff carbon filaments that adhere because of surface charge. E) Magnetic bead method. Water-cooled electromagnet (magnetic flux density gradient = 2400 gauss/m) was placed above plane of myocytes and pulled upwards on beads attached to β1 integrin by a monoclonal antibody. (Modified from ref. 9.) F) Effect on cationic and anionic amphiphiles on membrane curvature. Cationic amphiphiles collect in the inner leaflet and cause cupping, and anionic amphiphiles collect in the outer leaflet and cause a ball-shaped curvature (crenation).

face of the cell, and in some cases, a substantial fraction of the membrane is not stretched. Moreover, the membrane must adopt a cup-shape as it interacts with the curved glass tool or carbon fiber, and it is unclear how such membrane curvature complicates the nature of the applied stimulus.

An alternative method for studying mechanosensitive whole-cell currents is to press down on the myocyte with a glass microtool[42,91] or with the cantilever arm of an atomic-force microscope.[92] These maneuvers cause a localized and nonuniform stress by compressing the cell. As a depression forms at the site of contact, cytoplasm is displaced laterally, and the encompassing membrane is stretched. In a rod-shaped myocyte, the region of stretch would appear as a circumferential band. On the other hand, calculations indicate that pushing down on a rounded cultured chick myocyte by ~2-4 μm is sufficient to elicit a tension of 1-3 dyne/cm over most of the membrane bilayer and its attached cytoskeleton.[42] A combination of compression in the Z-axis and motion in the plane of the membrane (X-Y plane) also has been examined.[91]

Another approach is to stretch myocytes with magnetic beads coated-with monoclonal antibodies to integral membrane proteins.[93] This method applies force in an outward direction, along a line perpendicular to the sarcolemma. Finally, anionic and cationic amphiphiles that insert into the membrane have been used alter bilayer tension in cardiac myocytes,[94] as well as in heterologous expression systems.[48]

Cation SAC

In view of the distinct methods of stimulation, one might expect widely varying results. To the contrary, a fundamental observation is remarkably consistent across atrial, ventricular, and sinoatrial node myocytes: stretch activates a current that typically reverses between ~0 and -15 mV and has a nearly linear I-V relationship, as shown in Figure 5. Similar currents are reported in ventricular myocytes from guinea pig,[85,89] chick,[42] rat,[86,89] and mouse,[95] rat atrium,[56] rabbit sinoatrial node,[88] as well as human atria and ventricle.[89,90] In addition, a similar cation SAC currents are found in atrial and sinus nodal fibroblasts, and it is argued that fibroblast stretch affects cardiac electrical activity and heart rate (for review, see ref. 96).

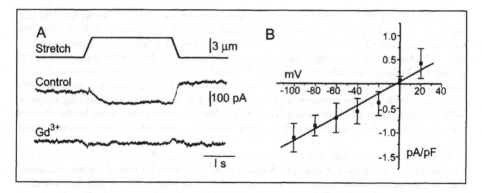

Figure 5. Whole-cell stretch activated currents from rat ventricular myocyte. A) Myocytes were attached between pipettes coated with covalently linked positively charged groups and stretched ~ 4 μm (top trace). Current under control recordings with physiological bathing media (middle trace) and after exposure to Gd^{3+} (lower trace) are shown. Gd^{3+} blocked the SAC. B) I-V relationship was linear and reversed at -6 mV. (Modified from ref. 86.)

As previously discussed for unitary SAC, whole-cell mechanosensitive currents are blocked by Gd^{3+},[42,86,90] *G. spatulata* venom,[42] and streptomycin.[73,97] An exception to this pharmacology is a cation SAC in rat atrium that is reported to be Gd^{3+}-insensitive.[56] On the other hand, these authors found Gd^{3+} blocked a stretch-insensitive background cation current.

Ion substitution experiments and an E_{rev} of ~0 to -15 mV establish that myocyte stretch activates a poorly selective cation current,[42,56,90] and that Ca^{2+} also is permeant.[89] It has been inferred from these data that only nonselective cation SAC are stimulated by whole-cell stretch. What happened to the K^+ SAC found at the single channel level? Hu and Sachs[42] argue that both a cation and K^+ SAC are activated by myocyte stretch and that both contribute to the recorded current. This conclusion is based on their observation of 90 pS K^+ SAC (E_{rev} = -70 mV) and 21 pS cation SAC (E_{rev} = -2 mV) unitary conductances that are turned on by pipette suction in the same preparation in which the whole-cell SAC E_{rev} is -16 mV. If both cation and K^+ SAC are activated by myocyte stretch, the intermediate E_{rev} could be obtained. The exact value of E_{rev} would reflect the relative number of K^+ and cation SAC, their unitary conductances, and open probabilities. A test of this notion would be to selectively block either the K^+ or cation SAC. This should shift E_{rev} for the whole-cell current, but a selective blocker has not been identified. Another test would be to differentially activate the two types of SAC. Bett and Sachs[92] reported in the same preparation that the cation SAC E_{rev} is identical with a series of graded stimuli that elicit graded whole-cell currents. If both K^+ and cation SAC are activated, as claimed,[42] their sensitivity to stretch must be identical to obtain a constant E_{rev} with graded activation,[92] a surprising notion. Additional work is needed to clarify the role of K^+ SAC in whole-cell currents.

Kinetics of the Whole-Cell Response

The time courses of activation and inactivation of mechanosensitive whole-cell currents were studied extensively by Sachs and colleagues.[42,91,92] Activation generally appears to be rapid (in the range of 10-100 ms), although detailed activation kinetics were not reported. Currents follow a 1 Hz sinusoidal stimulus without an obvious phase shift.[91] On the other hand, the response to applied stimulation can be complex. In some studies, the initially response is a spike of current lasting only ~10 ms, followed by a gradual increase in current over several seconds.[91] Moreover, Bett and Sachs[91] found that a stimulus that initially did not elicit a response generated a substantial response after 5 min of contact between the glass tool and the myocyte. Perhaps slow attachment of the membrane to the tool enhances transmission of force or activation of a signaling cascade is required to elicit the whole cell current.

Inactivation of whole cell SAC is noted in several preparations.[42,85,91,92] In chick ventricular myocytes, Hu and Sachs[42] described a rapid and substantial inactivation in 18% of cells. This was modeled as an O → I transition with a time constant of ~2 s. Myocytes seemed to recover in 10-15 s, but overall the current elicited by a second stretch 10-15 min latter was only ~75% of that by the initial stretch. Bett and Sachs[92] found ~50% of the current inactivated and a faster decay of current using a different stimulus protocol. This behavior was modeled with series and parallel springs representing elasticity and a parallel dashpot representing viscosity. Whereas rearrangement of the cytoskeleton may reduce the force experienced by the channel over time, intracellular signaling and autocrine molecules also might contribute to current decay.

It should be noted that not all investigators have detected inactivation of whole-cell stretch-induced currents. For example, Kamkin et al[89] applying local axial stretch concluded that the mechanosensitive current increases to a stable value within 200 ms and does not inactivate. The reason for this difference is unknown, but inactivation may depend on the characteristics of the stimulus.

Anion and Cation Currents with Integrin Stretch

To attempt to mimic the in situ situation, force can be applied specifically to molecules involved in physiologic force transmission. Browe and Baumgarten[93] attached magnetic beads coated with β1 integrin monoclonal antibodies to rabbit ventricular myocytes. When a magnetic coil was activated, the beads and attached integrins were pulled upward and perpendicular to the long axis of the myocyte and membrane plane (Fig. 4E). In solutions designed to isolate Cl⁻ currents, integrin stretch activates an outwardly rectifying Cl⁻ SAC that is blocked by tamoxifen, which also inhibits swelling-activated Cl⁻ channels, and PP2, which blocks src and FAK protein tyrosine kinases.[93] This Cl⁻ SAC shares many of the properties of the volume-sensitive Cl⁻ current, $I_{Cl,swell}$.[9] In physiologic solutions, integrin stretch also activates a linear cation current that reverses near -10 mV. Thus integrin stretch can stimulate both anion and cation currents, whereas other forms of myocyte stretch appear to elicit only cation currents.[63,97] Recent studies[98] demonstrated that activation of Cl⁻ SAC upon β1 integrin stretch depends on angiotensin II, AT_1 receptors, and the upregulation of sarcolemmal NADPH oxidase. NADPH oxidase produces superoxide anions that are rapidly converted to H_2O_2 by superoxide dismutase. The stretch-induced activation of Cl⁻ SAC is abrogated by losartan, an AT_1 blocker, DPI and AEBSF, inhibitors of NADPH oxidase, and catalase, which degrades H_2O_2 to H_2O. Moreover, exogenous H_2O_2 mimics the effect of stretch and activates a tamoxifen-sensitive outwardly rectifying Cl⁻ current.**

Although the magnetic bead method provides highly specific site of attachment via an antibody, the connection of integrin to the membrane and cytoskeleton and the interconnection of cytoskeletal proteins implies that forces will be distributed to multiple structural elements.[99] Applied forces are small (typically 1-10 pN/bead), and the resulting stretch is localized and nonuniform. Magnetic beads also can be coated with ECM proteins, as previously done, for example, to examine stretch-induced Ca^{2+} fluxes in fibroblasts[100] and other types of cells[101] or with antibodies to other putative components of the mechanotransduction process.

Amphiphile-Induced Membrane Curvature

Another approach for studying SAC is to insert charged amphiphiles into the membrane. Amphiphilic molecules possess both charged hydrophilic and uncharged hydrophobic domains and, therefore, have detergent-like properties. Because of the natural asymmetric distribution of negatively charged phospholipids in the inner and outer leaflets of the membrane bilayer, cationic amphiphiles, such as chlorpromazine, accumulate in the inner leaflet, whereas anionic amphiphiles, such as trinitrophenol and arachidonate, accumulate in the outer leaflet.[102] As a

---

** See *Notes Added in Proof.*

result, the tension in the bilayer is altered, and membrane curvature is induced (Fig. 4F). Cationic amphiphiles mimic the curvature caused by cell shrinkage (cup-shape), and anionic molecules mimic the curvature caused by cell swelling (ball-shaped) or crenation, as occurs in erythrocytes. Mechanosensitive exogenously expressed TREK-1 channels are activated by anionic amphiphiles, and their activation is reversed by cationic amphiphiles, responses attributed to alteration of bilayer tension.[48]

Amphiphiles have multiple effects in cardiac preparations. Anionic and cation amphiphiles appropriately mimic cell swelling and shrinkage, respectively, in regulating volume-sensitive Cl⁻ channels in canine ventricular myocytes in the absence of a volume change.[94] Anionic amphiphiles enhance I_{Ca-L} but inhibit I_{Ca-T} in rabbit ventricular myocytes, whereas cationic amphiphiles have the opposite effects.[103] Effects on Ca^{2+} currents were explain in terms of altered surface potential and altered lipid properties rather than as mechanosensitivity, however. Indeed, a plethora of cardiac membrane currents, transporters, and signaling cascades are modulated by amphiphiles, and the accumulation of certain amphiphiles during ischemia has been postulated as an important cause of arrhythmogenesis.[104] Because amphiphiles are not specific, caution is necessary before ascribing their action solely to altered membrane tension and mechanosensitivity.

Voltage- and Ligand-Gated Channels

Data on the effects of whole cell stretch on voltage-gated ion channels is somewhat contradictory, perhaps reflecting the details of the stimulus or its duration.

I_{K1}

Initial reports concluded that stretch did not alter I_{K1} in guinea-pig[85] or chick[105] ventricle. The response appears to depend on how the stimulus is delivered, however. Kamkin et al[89] found the Cs^+-sensitive difference current, I_{K1}, is enhanced by local stretch of guinea pig ventricular myocytes, but edgewise compression of myocytes rotated with their thin edge up causes significant inhibition of I_{K1}.[97] Inhibition of I_{K1} also is elicited by stretching β1 integrins.[93] Furthermore, the early study by Sasaki et al[85] showed a rectification of SAC current negative to -70 mV that is consistent with inhibition of I_{K1}, a possibility the authors briefly raised.

I_{Ca-L}

Initial reports also failed to identify consistent effects of stretch on I_{Ca-L} in guinea-pig, ferret, and rat ventricle.[85,106,107] By contrast, a 15-25% depression of I_{Ca-L} is observed in guinea pig, rat, mouse, and human ventricle using localized stretch.[89,90,95,97] Inhibition of I_{Ca-L} is prevented by dialysis with BAPTA, suggesting that Ca^{2+}-dependent inactivation of I_{Ca-L} following stretch-induced Ca^{2+} influx was responsible.

Other Channels and Transporters

There are a few reports regarding other channels and transporters in heart. No effect on I_{Na} is detected in chick ventricle.[105] Delayed rectifier K^+ channels are stimulated, inhibited[89,97] or unaffected,[85] depending on how stretch is applied. Prolonged stretch activates the Na^+-K^+ pump (dihydrouabain-sensitive current), presumably due to elevation of $[Na^+]_i$.[97] Because stretch-induced accumulation of $[Ca^{2+}]_i$ and $[Na^+]_i$ (see *infra*) has multiple effects on ion channels and transporters, both directly and via signaling cascades, it is likely that other effects will be identified in the future.

Effects of Stretch on Cardiac Electrical Activity

The overall effect of stretch on cardiac membrane currents is to produce an inward current at voltages negative to the plateau and an outward current at more positive potentials,[42,85,89] as also is seen when imposing an action potential waveform as the voltage command.[73] Such currents might be expected to depolarize resting potential and make the action potential more triangular, shortening action potential duration at 10 or 20% repolarization (e.g., APD_{10}) but prolonging APD_{75} or APD_{90}. The effect of the voltage trajectory on the activation and inactivation of voltage-dependent channels complicates predictions, however. Experimentally, stretch depolarizes the resting potential, decreases plateau amplitude, and either shortens APD or produces a crossover of the waveform as predicted above (for reviews, see refs. 3,108,109). The details of effects on APD depend, however, on the timing and duration of the stimulus and loading conditions. Nevertheless, the consistently observed abbreviation and negative shift of the plateau is expected to limit Ca^{2+} entry via both I_{Ca-L} and reverse mode Na^+-Ca^{2+} exchange. Because of the complexity of interactions, computer simulations are necessary to fully unravel how SAC activation affects various voltage-dependent pathways and the accumulation of ions. Significant progress has been made taking this approach,[88,109-113] and the consequences of nonuniform myocardial mechanics were considered recently.[114] In a complementary approach, stretch-activated currents calculated in real-time were applied to unstretched myocytes by current clamp and elicited electrical responses characteristic of stretch.[115]

Stretch also modulates cardiac rhythm. This was first recognized nearly a century ago as the Bainbridge effect, wherein increased right atrial filling causes an acceleration of heart rate that usually is attributed to autonomic reflexes. A mechanism intrinsic to the heart also is needed, however, because stretch enhances diastolic depolarization and automaticity of Purkinje fibers and the sinoatrial node,[116] and stretch-induced acceleration of pacemaker activity and depolarization of maximum diastolic potential are reproduced in isolated sinoatrial myocytes.[88] Computer modeling indicates activation of a linear cation SAC is sufficient to explain these data.[88]

Besides enhancing normal automaticity, stretch can elicit ectopic beats, tachycardia, and fibrillation.[3] Evidence for involvement of SAC in arrhythmias arising from brief stretch was first provided by Hansen et al,[117] who demonstrated in isolated canine hearts that increasing left ventricular volume for 50 ms elicits ectopy that is blocked by 1-10 μM Gd^{3+} but not by Ca^{2+} channel blockers. Graded volume pulses produced graded depolarizations in monophasic action potential recordings that were Gd^{3+}-sensitive and paralleled the propensity for arrhythmogenesis.[118] Gd^{3+} also blocks delayed afterdepolarizations in atria,[119] inhibits atrial fibrillation,[120] and minimizes dispersion of ventricular repolarization[121] during pressure overload. Because of the multiple effects of Gd^{3+}, a clearer case can be made with GsMTx4, the selective cation SAC blocker purified from tarantula venom.[77] GsMTx4 suppresses both atrial fibrillation induced by elevated intra-atrial pressure[78] and spontaneous depolarizations and runs of tachycardia in ventricular myocytes isolated from an aortic regurgitation model of heart failure.[122]

Stretch-Induced Elevation of $[Ca^{2+}]_i$ and $[Na^+]_i$

As discussed previously, electrophysiologic data indicate that cation SAC are permeant to Na^+ and Ca^{2+} and that Na^+ is the dominant inward charge carrier under physiologic conditions. This leads to the notion that activation of cation SAC will increase $[Na^+]_i$ and $[Ca^{2+}]_i$, and significant experimental support has been garnered for this idea. SAC do not act in isolation, however, and other transport processes including Na^+-Ca^{2+} and Na^+-H^+ exchange, Na^+-K^+ pump, I_{Ca-L}, sarcoplasmic reticulum (SR), and Ca^{2+} binding to troponin-c directly and indirectly contribute to the response to stretch (for review, see ref. 123).

A stretch-induced slow increase in Ca^{2+} transients over a number of minutes was first documented in intact cat papillary muscle and trabeculae based on aequorin bioluminescence.[124] Stretch also increases diastolic $[Ca^{2+}]_i$ measured with indo-1[125] or fura-2[106] in isolated guinea pig ventricular myocytes, and Ca^{2+} transients but not $[Ca^{2+}]_i$ in rat ventricular myocytes.[107] In a more intact model, Tavi et al[111,126] examined the response of rat left atrial appendage to physiological increases in intra-atrial pressure, 1-4 mm Hg, using indo-1. As shown for isolated rat myocytes, stretch increases Ca^{2+} transients.[111,126] In addition, diastolic $[Ca^{2+}]_i$ increases when the cytosol is made slightly acidotic (0.18 pH units).[126] Thus, a convincing case is made for stretch-induced Ca^{2+} uptake, but the source of Ca^{2+} is not defined by these studies.

Sigurdson et al[4] provided the first evidence for the involvement of SAC in stretch-induced Ca^{2+} uptake. On prodding fluo-3-loaded chick ventricular cells with a glass tool, they elicited localized Ca^{2+} influx that often induced waves of Ca^{2+}-induced Ca^{2+} release that spread throughout the myocyte. They also saw a diffuse increase in $[Ca^{2+}]_i$ by pulling on a neighboring attached myocyte. Importantly, they found that stretch-induced Ca^{2+} influx was blocked by 20 μM Gd^{3+}, suggesting SAC are involved, or by removing extracellular Ca^{2+}. Another SAC blocker, streptomycin (40 μM), prevents and reverses the large stretch-induced increase in $[Ca^{2+}]_i$ (up to 60% of the Ca^{2+} transient) on axial stretch of guinea pig myocytes[72,73] without altering I_{Ca-L},[72] or APD and contractility in unstretched myocytes.[73] Supporting the role of SAC, suppression of SR function with ryanodine and I_{Na} with tetrodotoxin has no effect; the Ca^{2+} channel blocker verapamil reduces stretch-induced Ca^{2+} entry by ~25% in unclamped myocytes, although this probably is due to altered action potential shape.[110] Ca^{2+} entry upon bidirectional stretch of neonatal rat myocytes cultured on laminin-[127] or collagen-coated[128] silicon rubber sheets is fully or 50% blocked by Gd^{3+}. Neither ryanodine[127] nor a combination or ruthenium red and procaine[128] suppress the rise in Ca^{2+} in this model. On the other hand, Ruwhof et al[128] found that diltiazem (100 μM) is almost as effective as Gd^{3+} in blocking the elevation of $[Ca^{2+}]_i$ with bidirectional stretch and fully blocks the response to prodding cells. Stretch failed to increase diastolic $[Ca^{2+}]_i$ in isolated adult rat ventricular myocytes[107] and trabeculae.[129] In fact a small decrease in diastolic $[Ca^{2+}]_i$ in rat ventricular trabeculae was reported by Alvarez et al,[130] although these authors simultaneously observed an increase in the Ca^{2+} transient.

Because Na^+ is the predominant charge carrier for inward cation SAC current, increased $[Na^+]_i$ also might be expected upon stretch. Such an increase in $[Na^+]_i$ is observed in rat ventricular trabeculae[130] and cat papillary muscle[131] but not in isolated rat ventricular myocytes,[107] as measured with the ratiometric fluorophore SBFI. A localized subsarcolemmal increase in $[Na^+]_i$ is described in isolated mouse ventricular myocytes by Isenberg et al,[97] who used the indicator sodium green with a pseudoratiometric method and simultaneously defined the sarcolemma by ANEPPS fluorescence. After a 4-min stretch, subsarcolemmal hot spots are noted; the most common pixel value in the hot spots corresponded to an increase of $[Na^+]_i$ from 11 to 24 mM, and 28% of the pixels gave values between 25 and 50 mM. An increase in cytoplasmic Na^+ content was confirmed by electron probe analysis,[97] which reports the sum of free and bound Na^+ (as much as 75% of cytoplasmic Na^+ is bound). Na^+ content increases 2.26-fold in the peripheral cytoplasm and 1.88-fold in the center of the cell after a 2-min stretch. Such large increases in $[Na^+]_i$ are a powerful stimulant for reverse mode Na^+-Ca^{2+} exchange, which mediates Ca^{2+} influx.

Although it is appealing to attribute these increases in Na^+ to Na^+ permeant cation SAC, direct evidence for the involvement of SAC has not been presented, and alternative mechanisms must be considered. Stretch evokes important autocrine/paracrine signaling cascades in heart. Stored angiotensin is rapidly released from myocytes[132] and, via AT_1 receptors, causes the secretion of endothelin-1.[133] In turn, binding of endothelin-1 to cardiac ET_A receptors activates the Na^+-H^+ exchanger (NHE1) by means of PKC, ERK1/2 and p90-RS kinase.[134,135] Stretch-induced stimulation of Na^+-H^+ exchange plays an important role in elevation of $[Na^+]_i$ and $[Ca^{2+}]_i$ in rat trabeculae[130] and cat papillary muscle,[131] and Na^+ accumulation is abrogated by AT_1 and ET_A receptor blockers. Na^+ entry via Na^+-H^+ exchange causes reverse mode Na^+-Ca^{2+} exchange that ultimately is dependent on activation of both ET_A[131] and AT_1[136] receptors.

Conclusion

The transduction of mechanical forces by SAC rapidly results in altered cardiac electrical activity and the influx of Ca^{2+} and Na^+. The accumulation of these ions contributes to altered contractile performance and may serve to initiate a variety of Ca^{2+}-dependent signaling cascades that regulate cardiac function via protein phosphorylation and ultimately gene expression. It is equally clear, however, that SAC are not the only mechanism for transducing mechanical stresses that impinge on the heart and that these signaling processes overlap. While the role of SAC in mechanoelectrical feedback is well established, much work and better tools will be needed to definitively identify the contribution of SAC to the initiation of signaling cascades.

Notes Added in Proof

*After submission of this manuscript, Hamill, Martinac and coworkers[137] reported that TRPC1 is a required component of stretch-activated, poorly selective, cation channels in vertebrate cells; TRPC1 can function as a homomeric channel, but it seems likely that it forms heteromers with TRPC3, TRPC4, TRPC5, and possibly polycystin-2. TRPC1 is express in human cardiac tissue based on RNA blots[138,139] and RT-PCR,[140] but it remains uncertain whether the transcripts arise in myocytes or other cardiac cells.

**Browe and Baumgarten[141] recently provided evidence that EGFR kinase and PI-3K also participate in the angiotensin-reactive oxygen signaling cascade that elicits Cl⁻ SAC and that Cl⁻ SAC is volume-sensitive. Cl⁻ SAC is inhibited by the EGFR kinase blocker AG 1478 and the PI-3K blockers wortmannin and LY 294002. Exogenous EGF elicits a similar tamoxifen-sensitive current, and both the stretch- and EGF-induced Cl⁻ currents are blocked by LY 294002, gp91ds-tat, a membrane-permeant peptide inhibitor of NADPH oxidase, and shrinkage in 1.5-times hyperosmotic bath solution.

Acknowledgements

Supported by grants from the National Institutes of Health (HL-26764, HL-65435) and the American Heart Association.

References

1. Tavi P, Laine M, Weckström M et al. Cardiac mechanotransduction: From sensing to disease and treatment. Trends Pharmacol Sci 2001; 22:254-260.
2. Kohl P, Ravens U (Guest eds). Focussed Issue: Mechano-Electrical Feedback and Cardiac Arrhythmias. Prog Biophys Mol Biol 2003; 82:1-266.
3. Franz MR. Mechano-electrical feedback in ventricular myocardium. Cardiovasc Res 1996; 32:15-24.
4. Sigurdson W, Ruknudin A, Sachs F. Calcium imaging of mechanically induced fluxes in tissue-cultured chick heart: Role of stretch-activated ion channels. Am J Physiol Heart Circ Physiol 1992; 262:H1110-H1115.
5. Sadoshima J, Izumo S. The cellular and molecular response of cardiac myocytes to mechanical stress. Annu Rev Physiol 1997; 59:551-571.
6. Kudoh S, Akazawa H, Takano H et al. Stretch-modulation of second messengers: Effects on cardiomyocyte ion transport. Prog Biophys Mol Biol 2003; 82:57-66.
7. Vandenberg JI, Rees SA, Wright AR et al. Cell swelling and ion transport pathways in cardiac myocytes. Cardiovasc Res 1996; 32:85-97.
8. Wright AR, Rees SA. Cardiac cell volume: Crystal clear or murky waters? A comparison with other cell types. Pharmacol Ther 1998; 80:89-121.
9. Baumgarten CM, Clemo HF. Swelling-activated chloride channels in cardiac physiology and pathophysiology. Prog Biophys Mol Biol 2003; 82:25-42.
10. Berthier C, Blaineau S. Supramolecular organization of the subsarcolemmal cytoskeleton of adult skeletal muscle fibers. A review. Biol Cell 1997; 89:413-434.
11. Clark KA, McElhinny AS, Beckerle MC et al. Striated muscle cytoarchitecture: An intricate web of form and function. Annu Rev Cell Dev Biol 2002; 18:637-706.
12. Burridge K, Chrzanowska-Wodnicka M. Focal adhesions, contractility, and signaling. Annu Rev Cell Dev Biol 1996; 12:463-518.
13. Defilippi P, Gismondi A, Santoni A et al. Signal transduction by integrins. Austin, TX: Landes Bioscience, 1997.
14. Borg TK, Goldsmith EC, Price R et al. Specialization at the Z line of cardiac myocytes. Cardiovasc Res 2000; 46:277-285.
15. Dalen H, Saetersdal T, Roli J et al. Effect of collagenase on surface expression of immunoreactive fibronectin and laminin in freshly isolated cardiac myocytes. J Mol Cell Cardiol 1998; 30:947-955.
16. Piper HM, Probst I, Schwartz P et al. Culturing of calcium stable adult cardiac myocytes. J Mol Cell Cardiol 1982; 14:397-412.
17. Sukharev SI, Blount P, Martinac B et al. A large-conductance mechanosensitive channel in E. coli encoded by mscL alone. Nature 1994; 368:265-268.
18. Hamill OP, Martinac B. Molecular basis of mechanotransduction in living cells. Physiol Rev 2001; 81:685-740.
19. Liu JD, Schrank B, Waterston RH. Interaction between a putative mechanosensory membrane channel and a collagen. Science 1996; 273:361-364.

20. Denker SP, Barber DL. Ion transport proteins anchor and regulate the cytoskeleton. Curr Opin Cell Biol 2002; 14:214-220.
21. Danowski BA, Imanaka-Yoshida K, Sanger JM et al. Costameres are sites of force transmission to the substratum in adult rat cardiomyocytes. J Cell Biol 1992; 118:1411-1420.
22. Wang N, Butler JP, Ingber DE. Mechanotransduction across the cell surface and through the cytoskeleton. Science 1993; 260:1124-1127.
23. Ross RS, Borg TK. Integrins and the myocardium. Circ Res 2001; 88:1112-1119.
24. Parsons JT. Focal adhesion kinase: The first ten years. J Cell Sci 2003; 116:1409-1416.
25. Rybakova IN, Patel JR, Ervasti JM. The dystrophin complex forms a mechanically strong link between the sarcolemma and costameric actin. J Cell Biol 2000; 150:1209-1214.
26. Straub V, Campbell KP. Muscular dystrophies and the dystrophin-glycoprotein complex. Curr Opin Neurol 1997; 10:168-175.
27. Franco-Obregon Jr A, Lansman JB. Mechanosensitive ion channels in skeletal muscle from normal and dystrophic mice. J Physiol (Lond) 1994; 481:299-309.
28. Franco-Obregon A, Lansman JB. Changes in mechanosensitive channel gating following mechanical stimulation in skeletal muscle myotubes from the mdx mouse. J Physiol 2002; 539:391-407.
29. Guharay F, Sachs F. Stretch-activated single ion channel currents in tissue-cultured embryonic chick skeletal muscle. J Physiol (Lond) 1984; 352:685-701.
30. Guharay F, Sachs F. Mechanotransducer ion channels in chick skeletal muscle: The effects of extracellular pH. J Physiol (Lond) 1985; 363:119-134.
31. Sachs F. Mechanical transduction in biological systems. Crit Rev Biomed Eng 1988; 16:141-169.
32. Sokabe M, Sachs F, Jing ZQ. Quantitative video microscopy of patch clamped membranes stress, strain, capacitance, and stretch channel activation. Biophys J 1991; 59:722-728.
33. Ruknudin A, Sachs F, Bustamante JO. Stretch-activated ion channels in tissue-cultured chick heart. Am J Physiol Heart Circ Physiol 1993; 264:H960-H972.
34. Sato R, Koumi S. Characterization of the stretch-activated chloride channel in isolated human atrial myocytes. J Membr Biol 1998; 163:67-76.
35. Kawakubo T, Naruse K, Matsubara T et al. Characterization of a newly found stretch-activated K$_{Ca,ATP}$ channel in cultured chick ventricular myocytes. Am J Physiol Heart Circ Physiol 1999; 276:H1827-H1838.
36. Levin KR, Page E. Quantitative studies on plasmalemmal folds and caveolae of rabbit ventricular myocardial cells. Circ Res 1980; 46:244-255.
37. Anderson RG. The caveolae membrane system. Annu Rev Biochem 1998; 67:199-225.
38. Kordylewski L, Goings GE, Page E. Rat atrial myocyte plasmalemmal caveolae in situ. Reversible experimental increases in caveolar size and in surface density of caveolar necks. Circ Res 1993; 73:135-146.
39. Kohl P, Cooper PJ, Holloway H. Effects of acute ventricular volume manipulation on in situ cardiomyocyte cell membrane configuration. Prog Biophys Mol Biol 2003; 82:221-227.
40. Brezden BL, Gardner DR, Morris CE. A potassium-selective channel in isolated Lymnaea stagnalis heart muscle cells. J Exp Biol 1986; 123:175-189.
41. Sigurdson WJ, Morris CE, Brezden BL et al. Stretch activation of a potassium channel in molluscan heart cells. J Exp Biol 1987; 127:191-210.
42. Hu H, Sachs F. Mechanically activated currents in chick heart cells. J Membr Biol 1996; 154:205-216.
43. Kim D. A mechanosensitive K$^+$ channel in heart cells. Activation by arachidonic acid. J Gen Physiol 1992; 100:1021-1040.
44. Niu W, Sachs F. Dynamic properties of stretch-activated K$^+$ channels in adult rat atrial myocytes. Prog Biophys Mol Biol 2003; 82:121-135.
45. Terrenoire C, Lauritzen I, Lesage F et al. A TREK-1-like potassium channel in atrial cells inhibited by beta-adrenergic stimulation and activated by volatile anesthetics. Circ Res 2001; 89:336-342.
46. Tan JH, Liu W, Saint DA. Trek-like potassium channels in rat cardiac ventricular myocytes are activated by intracellular ATP. J Membr Biol 2002; 185:201-207.
47. Lesage F, Lazdunski M. Molecular and functional properties of two-poredomain potassium channels. Am J Physiol Renal Physiol 2000; 279:F793-F801.
48. Patel AJ, Lazdunski M, Honore E. Lipid and mechano-gated 2P domain K$^+$ channels. Curr Opin Cell Biol 2001; 13:422-428.
49. O'Connell AD, Morton MJ, Hunter M. Two-pore domain K$^+$ channels-molecular sensors. Biochim Biophys Acta 2002; 1566:152-161.
50. Lesage F, Terrenoire C, Romey G et al. Human TREK2, a 2P domain mechano-sensitive K$^+$ channel with multiple regulations by polyunsaturated fatty acids, lysophospholipids, and G$_s$, G$_i$, and G$_q$ protein-coupled receptors. J Biol Chem 2000; 275:28398-28405.

51. Fink M, Duprat F, Lesage F et al. Cloning, functional expression and brain localization of a novel unconventional outward rectifier K⁺ channel. EMBO J 1996; 15:6854-6862.
52. Maingret F, Patel AJ, Lesage F et al. Mechano- or acid stimulation, two interactive modes of activation of the TREK-1 potassium channel. J Biol Chem 1999; 274:26691-26696.
53. Craelius W, Chen V, El-Sherif N. Stretch activated ion channels in ventricular myocytes. Biosci Rep 1988; 8:407-414.
54. Sadoshima J, Takahashi T, Jahn L et al. Roles of mechano-sensitive ion channels, cytoskeleton, and contractile activity in stretch-induced immediate-early gene expression and hypertrophy of cardiac myocytes. Proc Natl Acad Sci USA 1992; 89:9905-9909.
55. Kim D. Novel cation-selective mechanosensitive ion channel in the atrial cell membrane. Circ Res 1993; 72:225-231.
56. Zhang YH, Youm JB, Sung HK et al. Stretch-activated and background nonselective cation channels in rat atrial myocytes. J Physiol (Lond) 2000; 523:607-619.
57. Bustamante JO, Ruknudin A, Sachs F. Stretch-activated channels in heart cells: Relevance to cardiac hypertrophy. J Cardiovasc Pharmacol 1991; 17(Suppl 2):S110-S113.
58. Hoyer J, Distler A, Haase W et al. Ca²⁺ influx through stretch-activated cation channels activates maxi K⁺ channels in porcine endocardial endothelium. Proc Natl Acad Sci USA 1994; 91:2367-2371.
59. Vennekens R, Voets T, Bindels RJ et al. Current understanding of mammalian TRP homologues. Cell Calcium 2002; 31:253-264.
60. Mutai H, Heller S. Vertebrate and invertebrate TRPV-like mechanoreceptors. Cell Calcium 2003; 33:471-478.
61. Liedtke W, Choe Y, Marti-Renom MA et al. Vanilloid receptor-related osmotically activated channel (VR-OAC), a candidate vertebrate osmoreceptor. Cell 2000; 103:525-535.
62. Yang XC, Sachs F. Block of stretch-activated ion channels in Xenopus oocytes by gadolinium and calcium ions. Science 1989; 243:1068-1071.
63. Hu H, Sachs F. Stretch-activated ion channels in the heart. J Mol Cell Cardiol 1997; 29:1511-1523.
64. Maingret F, Fosset M, Lesage F et al. TRAAK is a mammalian neuronal mechano-gated K⁺ channel. J Biol Chem 1999; 274:1381-1387.
65. Maingret F, Patel AJ, Lesage F et al. Lysophospholipids open the two-pore domain mechano-gated K⁺ channels TREK-1 and TRAAK. J Biol Chem 2000; 275:10128-10133.
66. Li GR, Baumgarten CM. Modulation of cardiac Na⁺ current by gadolinium, a blocker of stretch-induced arrhythmias. Am J Physiol Heart Circ Physiol 2001; 280:H272-H279.
67. Lacampagne A, Gannier F, Argibay J et al. The stretch-activated ion channel blocker gadolinium also blocks L-type calcium channels in isolated ventricular myocytes of the guinea-pig. Biochim Biophys Acta 1994; 1191:205-208.
68. Pascarel C, Hongo K, Cazorla O et al. Different effects of gadolinium on I_Kr, I_Ks and I_K1 in guinea-pig isolated ventricular myocytes. Br J Pharmacol 1998; 124:356-360.
69. Zhang YH, Hancox JC. Gadolinium inhibits Na⁺-Ca²⁺ exchanger current in guinea-pig isolated ventricular myocytes. Br J Pharmacol 2000; 130:485-488.
70. Caldwell RA, Clemo HF, Baumgarten CM. Using gadolinium to identify stretch-activated channels: Technical considerations. Am J Physiol Cell Physiol 1998; 275:C619-C621.
71. Hamill OP, McBride Jr DW. The pharmacology of mechanogated membrane ion channels. Pharmacol Rev 1996; 48:231-252.
72. Gannier F, White E, Lacampagne A et al. Streptomycin reverses a large stretch induced increases in [Ca²⁺]ᵢ in isolated guinea pig ventricular myocytes. Cardiovasc Res 1994; 28:1193-1198.
73. Belus A, White E. Streptomycin and intracellular calcium modulate the response of single guinea-pig ventricular myocytes to axial stretch. J Physiol (Lond) 2003; 546:501-509.
74. Belus A, White E. Effects of streptomycin sulphate on I_Ca-L, I_Kr and I_Ks in guinea-pig ventricular myocytes. Eur J Pharmacol 2002; 445:171-178.
75. Pérez-Miles F, Lucas SM, da Silva Jr PI et al. Systematic revision and cladistic analysis of Theraphosinae (Araneae: Theraphosidae). Mygalomorph 1996; 1:33-68.
76. Chen Y, Simasko SM, Niggel J et al. Ca²⁺ uptake in GH₃ cells during hypotonic swelling: The sensory role of stretch-activated ion channels. Am J Physiol Cell Physiol 1996; 270:C1790-C1798.
77. Suchyna TM, Johnson JH, Hamer K et al. Identification of a peptide toxin from Grammostola spatulata spider venom that blocks cation-selective stretch-activated channels. J Gen Physiol 2000; 115:583-598.
78. Bode F, Sachs F, Franz MR. Tarantula peptide inhibits atrial fibrillation. Nature 2001; 409:35-36.
79. Christensen M, Strange K. Developmental regulation of a novel outwardly rectifying mechanosensitive anion channel in Caenorhabditis elegans. J Biol Chem 2001; 276:45024-45030.
80. Van Wagoner DR. Mechanosensitive gating of atrial ATP-sensitive potassium channels. Circ Res 1993; 72:973-983.

81. Van Wagoner DR, Lamorgese M. Ischemia potentiates the mechanosensitive modulation of atrial ATP-sensitive potassium channels. Ann NY Acad Sci 1994; 723:392-395.
82. Pleumsamran A, Kim D. Membrane stretch augments the cardiac muscarinic K⁺ channel activity. J Membr Biol 1995; 148:287-297.
83. Zile MR, Cowles MK, Buckley JM et al. Gel stretch method: A new method to measure constitutive properties of cardiac muscle cells. Am J Physiol Heart Circ Physiol 1998; 274:H2188-H2202.
84. Cazorla O, Pascarel C, Brette F et al. Modulation of ions channels and membrane receptors activities by mechanical interventions in cardiomyocytes: Possible mechanisms for mechanosensitivity. Prog Biophys Molec Biol 1999; 71:29-58.
85. Sasaki N, Mitsuiye T, Noma A. Effects of mechanical stretch on membrane currents of single ventricular myocytes of guinea-pig heart. Jpn J Physiol 1992; 42:957-970.
86. Zeng T, Bett GC, Sachs F. Stretch-activated whole cell currents in adult rat cardiac myocytes. Am J Physiol Heart Circ Physiol 2000; 278:H548-H557.
87. Le Guennec JY, Peineau N, Argibay JA et al. A new method of attachment of isolated mammalian ventricular myocytes for tension recording: Length dependence of passive and active tension. J Mol Cell Cardiol 1990; 22:1083-1093.
88. Cooper PJ, Lei M, Cheng LX et al. Selected contribution: Axial stretch increases spontaneous pacemaker activity in rabbit isolated sinoatrial node cells. J Appl Physiol 2000; 89:2099-2104.
89. Kamkin A, Kiseleva I, Isenberg G. Stretch-activated currents in ventricular myocytes: Amplitude and arrhythmogenic effects increase with hypertrophy. Cardiovasc Res 2000; 48:409-420.
90. Kamkin A, Kiseleva I, Wagner KD et al. Characterization of stretch-activated ion currents in isolated atrial myocytes from human hearts. Pflugers Arch 2003; 446:339-346.
91. Bett GC, Sachs F. Whole-cell mechanosensitive currents in rat ventricular myocytes activated by direct stimulation. J Membr Biol 2000; 173:255-263.
92. Bett GC, Sachs F. Activation and inactivation of mechanosensitive currents in the chick heart. J Membr Biol 2000; 173:237-254.
93. Browe DM, Baumgarten CM. Stretch of β1 integrin activates an outwardly-rectifying chloride current via FAK and Src in rabbit ventricular myocytes. J Gen Physiol 2003; 122:689-702.
94. Tseng GN. Cell swelling increases membrane conductance of canine cardiac cells: Evidence for a volume-sensitive Cl channel. Am J Physiol Cell Physiol 1992; 262:C1056-C1068.
95. Kamkin A, Kiseleva I, Isenberg G. Ion selectivity of stretch-activated cation currents in mouse ventricular myocytes. Pflugers Arch 2003; 446:220-231.
96. Kamkin A, Kiseleva I, Isenberg G et al. Cardiac fibroblasts and the mechano-electric feedback mechanism in healthy and diseased hearts. Prog Biophys Mol Biol 2003; 82:111-120.
97. Isenberg G, Kazanski V, Kondratev D et al. Differential effects of stretch and compression on membrane currents and [Na⁺]c in ventricular myocytes. Prog Biophys Mol Biol 2003; 82:43-56.
98. Browe DM, Baumgarten CM. Angiotensin II (AT1) receptors and NADPH oxides regulate Cl⁻ SAC elicited by β1 integrin stretch in rabbit ventricular myocytes. J Gen Physiol 2004; 124:273-287.
99. Ingber DE. Tensegrity: The architectural basis of cellular mechanotransduction. Annu Rev Physiol 1997; 59:575-599.
100. Glogauer M, Ferrier J, McCulloch CAG. Magnetic fields applied to collagen-coated ferric oxide beads induce stretch-activated Ca²⁺ flux in fibroblasts. Am J Physiol Cell Physiol 1995; 269:C1093-C1104.
101. Niggel J, Sigurdson W, Sachs F. Mechanically induced calcium movements in astrocytes, bovine aortic endothelial cells and C₆ glioma cells. J Membr Biol 2000; 174:121-134.
102. Sheetz MP, Singer SJ. Biological membranes as bilayer couples. A molecular mechanism of drug-erythrocyte interactions. Proc Natl Acad Sci USA 1974; 71:4457-4461.
103. Post JA, Ji S, Leonards KS et al. Effects of charged amphiphiles on cardiac cell contractility are mediated via effects on Ca²⁺ current. Am J Physiol Heart Circ Physiol 1991; 260:H759-H769.
104. Corr PB, McHowat J, Yan G et al. Lipid-derived amphiphiles and their contribution to arrhythmogenesis during myocardial ischemia. In: Sperelakis N, ed. Physiology and Pathophysiology of the Heart.Boston, MA: Kluwer Academic Publishers, 1995:527-545.
105. Hu H, Sachs F. Single-channel and whole-cell studies of mechanosensitive currents in chick heart. Biophys J 1996; 70:A347.
106. White E. The lack of effect of increasing cell length on L-type clcium current in isolated ferret ventricular myocytes. J Physiol (Lond) 1995; 483:P13.
107. Hongo K, White E, Le Guennec JY et al. Changes in [Ca²⁺]i, [Na⁺]i and Ca²⁺ current in isolated rat ventricular myocytes following an increase in cell length. J Physiol (Lond) 1996; 491:609-619.
108. Zabel M, Koller BS, Sachs F et al. Stretch-induced voltage changes in the isolated beating heart: Importance of the timing of stretch and implications for stretch- activated ion channels. Cardiovasc Res 1996; 32:120-130.

109. Kohl P, Hunter P, Noble D. Stretch-induced changes in heart rate and rhythm: Clinical observations, experiments and mathematical models. Prog Biophys Molec Biol 1999; 71:91-138.
110. Gannier F, White E, Garnier D et al. A possible mechanism for large stretch-induced increase in [Ca^{2+}]$_i$ in isolated guinea-pig ventricular myocytes. Cardiovasc Res 1996; 32:158-167.
111. Tavi P, Han C, Weckström M. Mechanisms of stretch-induced changes in [Ca^{2+}]$_i$ in rat atrial myocytes: Role of increased troponin C affinity and stretch-activated ion channels. Circ Res 1998; 83:1165-1177.
112. Kohl P, Day K, Noble D. Cellular mechanisms of cardiac mechano-electric feedback in a mathematical model. Can J Cardiol 1998; 14:111-119.
113. Knudsen Z, Holden AV, Brindley J. Qualitative modeling of mechanoelectrical feedback in a ventricular cell. Bull Math Biol 1997; 59:1155-1181.
114. Markhasin VS, Solovyova O, Katsnelson LB et al. Mechano-electric interactions in heterogeneous myocardium: Development of fundamental experimental and theoretical models. Prog Biophys Mol Biol 2003; 82:207-220.
115. Wagner MB, Kumar R, Joyner RW et al. Induced automaticity in isolated rat atrial cells by incorporation of a stretch-activaed conductance. Pflugers Arch 2004; 447:819-829.
116. Hoffman BF, Cranefield PF. Electrophysiology of the Heart. New York: McGraw-Hill, 1960:117,197-198.
117. Hansen DE, Borganelli M, Stacy Jr GP et al. Dose-dependent inhibition of stretch-induced arrhythmias by gadolinium in isolated canine ventricles. Evidence for a unique mode of antiarrhythmic action. Circ Res 1991; 69:820-831.
118. Stacy Jr GP, Jobe RL, Taylor LK et al. Stretch-induced depolarizations as a trigger of arrhythmias in isolated canine left ventricles. Am J Physiol Heart Circ Physiol 1992; 263:H613-H621.
119. Tavi P, Laine M, Weckström M. Effect of gadolinium on stretch-induced changes in contraction and intracellularly recorded action- and afterpotentials of rat isolated atrium. Br J Pharmacol 1996; 118:407-413.
120. Bode F, Katchman A, Woosley RL et al. Gadolinium decreases stretch-induced vulnerability to atrial fibrillation. Circulat 2000; 101:2200-2205.
121. Takagi S, Miyazaki T, Moritani K et al. Gadolinium suppresses stretch-induced increases in the differences in epicardial and endocardial monophasic action potential durations and ventricular arrhythmias in dogs. Jpn Circ J 1999; 63:296-302.
122. Clemo HF, Baumgarten CM. Cation stretch-activated channels cause spontaneous depolarizations in aortic regurgitation-induced heart failure. J Physiol (Lond) 2002; 544P:24S.
123. Calaghan SC, Belus A, White E. Do stretch-induced changes in intracellular calcium modify the electrical activity of cardiac muscle? Prog Biophys Mol Biol 2003; 82:81-95.
124. Allen DG, Kurihara S. The effects of muscle length on intracellular calcium transients in mammalian cardiac muscle. J Physiol 1982; 327:79-94.
125. Le Guennec JY, White E, Gannier F et al. Stretch-induced increase of resting intracellular calcium concentration in single guinea-pig ventricular myocytes. Exp Physiol 1991; 76:975-978.
126. Tavi P, Han C, Weckström M. Intracellular acidosis modulates the stretch-induced changes in E-C coupling of the rat atrium. Acta Physiol Scand 1999; 167:203-213.
127. Tatsukawa Y, Kiyosue T, Arita M. Mechanical stretch increases intracellular calcium concentration in cultured ventricular cells from neonatal rats. Heart Vessels 1997; 12:128-135.
128. Ruwhof C, van Wamel JT, Noordzij LA et al. Mechanical stress stimulates phospholipase C activity and intracellular calcium ion levels in neonatal rat cardiomyocytes. Cell Calcium 2001; 29:73-83.
129. Kentish JC, Wrzosek A. Changes in force and cytosolic Ca^{2+} concentration after length changes in isolated rat ventricular trabeculae. J Physiol (Lond) 1998; 506:431-444.
130. Alvarez BV, Perez NG, Ennis IL et al. Mechanisms underlying the increase in force and Ca^{2+} transient that follow stretch of cardiac muscle: A possible explanation of the Anrep effect. Circ Res 1999; 85:716-722.
131. Perez NG, de Hurtado MC, Cingolani HE. Reverse mode of the Na$^+$-Ca^{2+} exchange after myocardial stretch: Underlying mechanism of the slow force response. Circ Res 2001; 88:376-382.
132. Sadoshima J, Xu Y, Slayter HS et al. Autocrine release of angiotensin II mediates stretch-induced hypertrophy of cardiac myocytes in vitro. Cell 1993; 75:977-984.
133. Ito H, Hirata Y, Adachi S et al. Endothelin-1 is an autocrine/paracrine factor in the mechanism of angiotensin II-induced hypertrophy in cultured rat cardiomyocytes. J Clin Invest 1993; 92:398-403.
134. Yamazaki T, Komuro I, Kudoh S et al. Role of ion channels and exchangers in mechanical stretch-induced cardiomyocyte hypertrophy. Circ Res 1998; 82:430-437.
135. Sugden PH. An overview of endothelin signaling in the cardiac myocyte. J Mol Cell Cardiol 2003; 35:871-886.

136. Aiello EA, Villa-Abrille MC, Cingolani HE. Autocrine stimulation of cardiac Na$^+$-Ca^{2+} exchanger currents by endogenous endothelin released by angiotensin II. Circ Res 2002; 90:374-376.
137. Maroto R, Raso A, Wood TG et al. TRPC1 forms the stretch-activated cation channel in vertebrate cells. Nat Cell Biol 2005; 7:179-185.
138. Wes PD, Chevesich J, Jeromin et al. A TRPC1, a human homolog of a Drosophilia store- operated channel. Proc Natl Acad Sci USA 1995; 92:9652-9656.
139. Zhu X, Chu PB, Peyton M et al. Molecular cloning of a widely expressed human homologue for the Drosophila trp gene. FEBS Lett 1995; 373:193-198.
140. Riccio A, Medhurst AD, Mattei C et al. mRNA distribution analysis of human TRPC family in CNS and peripheral tissues. Molec Brain Res 2002; 109:95-104.
141. Browe DM, Baumgarten CM. EGFR kinase regulates volume-sensitive chloride current elicited by integrin stretch via PI-3K and NADPH oxidase in ventricular myocytes. J Gen Physiol 1006; 127:237-251.

CHAPTER 3

The Role of the Sarcomere and Cytoskeleton in Cardiac Mechanotransduction

Sarah C. Calaghan and Ed White*

Abstract

The basic contractile unit of the cardiac myocyte is the sarcomere. Force develops as a result of the interaction of myosin heads with the actin thin filament. Actin filaments are directly connected to the Z line of the sarcomere, whereas myosin filaments are secured via the giant elastic protein titin. When cardiac muscle is stretched there is an immediate increase in contractility. This is an acute and fundamental cardiac adaptive response to an increase in demand. Evidence suggests that an increase in the probability of crossbridge formation, through titin strain and positive cooperative mechanisms, underlies the length-dependent activation of cardiac muscle. The sarcomere is connected to the sarcolemma by cytoskeletal components which link the Z-line with the membrane-spanning integrins and dystroglycan complex. Integrins and dystroglycan, in turn, bind to components of the extracellular matrix, such as laminin, which sheath the cardiac myocyte. Connections also exist between Z-line and nucleus via the intermediate filament protein desmin. The intracellular connections between the Z-line of the sarcomere and the sarcolemma allow transmission of force developed by the myofilaments. However, the physical pathway that links the extracellular matrix, membrane-spanning proteins, and the cell interior also plays a fundamental role in mechanotransduction. These links allow the cell to sense and respond to mechanical stimuli through connections with the cytoskeleton and activation of signalling cascades.

Introduction

The basic contractile unit of the cardiac myocyte is the sarcomere. Force develops as a result of the interaction of myosin heads with the actin thin filament. Actin filaments are directly connected to the Z line of the sarcomere, whereas myosin filaments are secured via the giant elastic protein titin (see Fig. 1). Force developed within the sarcomere is transmitted both longitudinally and laterally across the sarcolemma of the myocyte. Lateral connections of the Z-line of peripheral myofibrils to the sarcolemma are made via membrane spanning integrins and the dystroglycan complex, which in turn interact with components of the extracellular matrix that sheath the cardiac myocyte. Longitudinal transmission takes place at the interdigitated ends of the myocyte at specialised junctions known as fascia adherens and desmosomes which form cell-to-cell connections of actin and desmin intermediate filaments respectively (see Fig. 1). However, the intracellular connections between sarcomere and sarcolemma are not only important for transmission of force developed within the sarcomere, they also play a vital role in mechanotransduction. Increased filling of the ventricles, for example during exercise,

*Corresponding Author: Ed White—School of Biomedical Sciences, University of Leeds, Leeds, LS2 9JT, U.K. Email: e.white@leeds.ac.uk

Cardiac Mechanotransduction, edited by Matti Weckström and Pasi Tavi.
©2007 Landes Bioscience and Springer Science+Business Media.

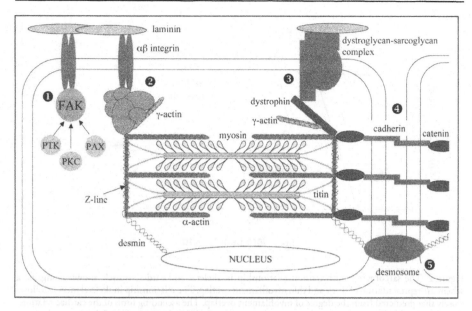

Figure 1. A simple schematic representation of some physical connections that provide a pathway for mechanotransduction within the cardiac myocyte. The extracellular matrix (ECM) protein laminin binds with 2 membrane-spanning proteins, the αβ integrins and dystroglycan. Both integrins and dystroglycan are concentrated in specialised regions of the sarcolemma (costameres) which line up with the Z band of peripheral myofibrils. ❶ Association of integrins with ECM components causes clustering of integrins which activates focal adhesion kinase (FAK) via autophosphorylation. This, in turn, leads to the association and activation of signalling molecules including other protein tyrosine kinases (PTK), protein kinase C (PKC), and paxillin (PAX). ❷ Integrins are also found in association with cytoskeletal proteins including vinculin, talin, spectrin, ankyrin, α-actinin. These components make connections to α-actin at the Z-line of the sarcomere, and with cytoskeletal γ-actin (either in a sub-membranous cortical lattice or in actin microfilament bundles). Desmin intermediate filaments connect the sarcolemma and Z-line via spectrin and ankyrin. Desmin also links Z-lines in adjacent sarcomeres, and forms a potential mechanotransductive pathway between the Z-line and nucleus. ❸ The dystroglycan/sarcoglycan complex interacts with cortical γ-actin and sarcomeric α-actin via dystrophin. Dystrophin may also link desmin filaments to the sarcolemma. As well as forming mechanotransductive pathways, these lateral connections from the Z-line to the sarcolemma are vital for force transmission from the contractile machinery. Mechanical connections between cells, vital for longitudinal force transmission, are made at cell-to-cell junctions via actin at fascia adherens (❹), and via intermediate filaments at desmosomes (❺).

causes stretch of the myocardium, and for the heart to function normally it must be able to sense and respond to this mechanical stimulus. Myocardial stretch increases the level of activation of the contractile machinery, and results in a biphasic increase in the force with which the heart contracts. In the long term, however, stretch can trigger the activation of intracellular signalling pathways that lead to the development of cardiac hypertrophy, and ultimately heart failure. In this chapter we will deal initially with the mechanisms, at the level of the sarcomere, that underlie length-dependent activation of cardiac muscle. We will then consider the role of the cardiac cell cytoskeleton and its extracellular connections in the mechanotransductive process.

The Sarcomere and Force Development

Stretch and the Rapid Increase in Active Force

Perhaps the most famous response of the myocardium to mechanical stimulation is summarised by the Frank-Starling law of the heart, which states that the output of the heart is

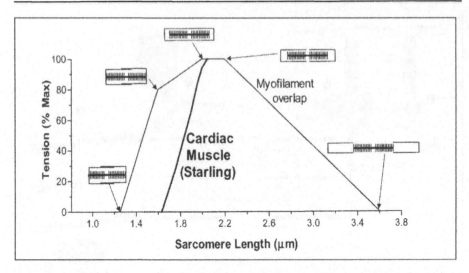

Figure 2. The sarcomere length-tension relationship in skeletal and cardiac muscle shown alongside a schematic representation of thick and thin myofilament overlap. The relationship in skeletal muscle follows closely that predicted from the degree of myofilament overlap. The ascending limb of the cardiac relationship (shown in bold) is much steeper between sarcomere lengths (SL) of 1.8-2.0 μm and shows some increase in tension over the SL range 2.0-2.2 μm when overlap is constant and optimal (from Bers).[134]

equal to, and determined by, the amount of blood flowing into the heart.[1,2] What this means in essence is that stretch of the myocardium increases its contractility. The importance of this response is evident when one considers the increase in stroke volume that occurs during exercise, when there is increased ventricular filling and thus dilation, or stretch, of the ventricles. The increase in contractility is concurrent with an increase in heart rate partially brought about by a dilation-dependent quickening of the intrinsic pacing frequency of the sino-atrial node (the Bainbridge effect).[3]

The increase in contractility, or active force, seen upon stretch is related to the change in the length of the individual sarcomeres, measured from Z-line to Z-line, within the ventricular myocyte. When a cardiac myocyte is stretched there is a steep increase in force from resting sarcomere length (SL) of approximately 1.8 μm up to a maximum at a length (L_{max}) of ≈ 2.2-2.3 μm. At longer lengths active force declines.

Myofilament Overlap

Classic experiments in skeletal muscle from Gordon et al explained the length-tension relationship in terms of thick and thin myofilament overlap (Fig. 2).[4] The degree of overlap of the myofilaments determines the number of crossbridges that can potentially cycle. As muscle is stretched from resting SL, there is a progressive decrease in double thin filament overlap until an optimum overlap of thick and thin filaments is reached at SLs between 2.0-2.2 μm. At longer lengths, tension falls as myofilament overlap declines and with it the opportunity for crossbridge formation. However, in cardiac muscle myofilament overlap does not fully account for the length-tension relationship. Between SLs of 1.8 and 2.0 μm, the relationship between length and tension is much steeper than that for skeletal muscle,[5] and there can be a marked length-dependent increase in tension over the SL range 2.0-2.2 μm when the number of cross-bridges that can potentially form remains constant.[6,7] It is now thought that the degree of activation of cardiac muscle changes with length and it is this, rather than myofilament overlap, that makes the major contribution to the length-tension relationship (see reviews by Allen and Kentish and Lakatta).[7,8]

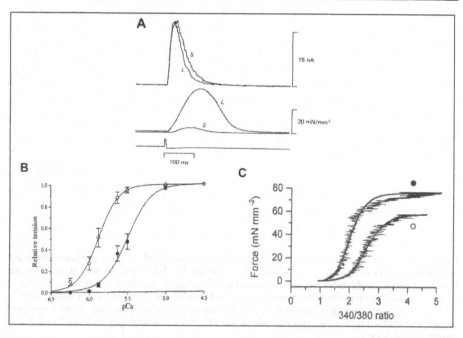

Figure 3. Stretch increases myofilament Ca^{2+} sensitivity. A) [Ca^{2+}]$_i$ transients (upper traces, measured by aqueorin) and tension (lower traces) in an intact rat trabecula at 100% (L) and 81% (S) of L$_{max}$. Stretch increases force in the absence of an increase in the amplitude of the [Ca^{2+}]$_i$ transient (From Allen and Kurihara).[9] B) The tension-pCa curve is shifted to the left, and pCa$_{50}$ (pCa at half maximal tension) is increased in skinned rat ventricular myocytes stretched from a SL of 1.9 μm (●) to 2.3 μm (0). Tension is normalised to maximal tension at each SL (from Cazorla et al).[37] C) The relationship between force and [Ca^{2+}]$_i$ (given as fura-2 fluorescence ratio) in an intact rat ventricular trabecula is measured during relaxation from a tetanic contracture induced by stimulation at 10 Hz in the presence of 8 mM [Ca^{2+}]$_o$, 1 μM ryanodine, and 30 μM cyclopiazonic acid, at approximately 95% (●) and 86% (0) L$_{max}$. A leftwards shift in the relationship and an increase in maximal tetanic force can be seen at the longer muscle length (from Kentish and Wrzosek).[12]

Myofilament Ca^{2+} Sensitivity

Allen and Kurihara showed that the increase in force seen immediately upon stretch, in both rat and cat papillary muscles, is not associated with an increase in the magnitude of the intracellular [Ca^{2+}] ([Ca^{2+}]$_i$) transient (Fig. 3A).[9] If the supply of Ca^{2+} to the myofilaments is not length-dependent, activation must occur by an increase in the sensitivity of the myofilaments to Ca^{2+} An increase in myofilament Ca^{2+} sensitivity describes the situation in which a given concentration of (sub-maximal) Ca^{2+} gives a greater degree of activation, and this is reflected in a leftward shift of the tension-pCa relationship (Fig. 3B). Changes in Ca^{2+} sensitivity are often indexed by the change in the Ca^{2+} concentration required for half-maximal activation ([Ca^{2+}]$_{50}$), or the pCa required for half maximal activation (pCa$_{50}$). There is substantial evidence for a length-dependent shift in myofilament Ca^{2+} sensitivity in cardiac muscle (see reviews by Allen and Kentish, Lakatta, and Calaghan and White).[7,8,10] Using skinned rat trabeculae, Hibberd and Jewell described a leftwards shift in the tension-pCa relationship with increasing sarcomere length, and a decrease in the [Ca^{2+}]$_{50}$.[11] A leftwards shift in the force-[Ca^{2+}]$_i$ relationship at longer lengths has also been reported in intact rat trabeculae during relaxation from tetanic stimulation, when [Ca^{2+}] was equated to fura-2 fluorescence (Fig. 3C).[12] The affinity of the thin filament Ca^{2+}-binding protein troponin C (TnC) for Ca^{2+} is one way in which myofilament Ca^{2+} sensitivity is modulated. Indeed the length-tension relationship in

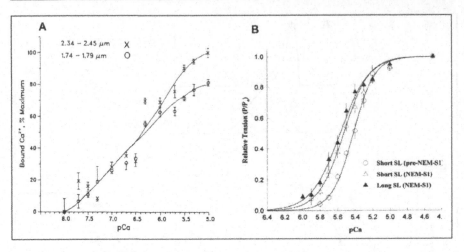

Figure 4. A) Stretch increases the amount of Ca²⁺ bound to troponin C (TnC). Experiments were performed at long length (SLs of 2.34-2.45 μm; X) and shorter length (1.74-1.79 μm; o) with skinned bovine ventricular muscle in which TnC was the only major Ca²⁺-binding protein (from Hofmann and Fuchs).[13] B) A strong-binding derivative of the myosin head (N-ethylmaleimide-modified myosin subfragment 1; NEM-S1) eliminates the length-dependent shift in the tension-pCa relationship. Experiments were performed with skinned rat cardiac myocytes held at short SL (≈1.9 μm) and long SL (≈2.25 μm) (from Fitzsimons and Moss).[23]

cardiac muscle is known to involve changes in TnC affinity for Ca²⁺ and a length-dependent change in the binding of Ca²⁺ by TnC has been shown (Fig. 4A).[13] It has been proposed that the length-dependent increase in Ca²⁺ sensitivity of cardiac muscle is dependent on intrinsic properties of the cardiac isoform of TnC,[14-16] although this is not universally accepted.[17,18] Furthermore, the idea that changes in TnC affinity for Ca²⁺ are the central player in length-dependent activation of cardiac muscle has been brought into question as length-dependence occurs even in the absence of Ca²⁺ for example, when activation is induced by reducing the concentration of ATP.[19]

Factors Affecting Length-Dependent Increases in Myofilament Ca²⁺ Sensitivity

The term length-tension relation is something of a misnomer as there is good evidence that it is force, rather than length per se, that modulates myofilament Ca²⁺ sensitivity upon stretch. Allen and Kentish showed that when Ca²⁺-activated skinned muscle was allowed to shorten, the amount of Ca²⁺ liberated from the myofilaments depended more upon the change in force than the change in length.[20] Indeed, the force-producing crossbridge has been assigned a central role in the length-tension relationship of cardiac muscle, illustrated, for example, by the observation that length-dependent changes in the Ca²⁺ binding of TnC can be abolished using vanadate to prevent crossbridge formation.[13]

One process that has been proposed to play a role in the Frank-Starling mechanism in cardiac muscle is the positive cooperativity of crossbridge binding. This can be understood by reference to the three-state model of thin filament activation,[21] based on experiments by Bremel and Weber.[22] In the absence of Ca²⁺ the thin filament protein tropomyosin blocks the interaction of actin and myosin. When Ca²⁺ binds to TnC there is some movement of tropomyosin into the groove between the actin filaments which allows the myosin heads to bind weakly to actin. The weakly-bound crossbridge isomerises into a strongly bound crossbridge which is responsible for force generation, but also shifts tropomyosin further into the groove between the actin filaments allowing recruitment of more crossbridges. Positive cooperativity describes this situation in which the formation of a strong-binding crossbridge makes the formation of another more likely. The importance of the strong-binding crossbridge in the Frank-Starling mechanism has been demonstrated by Fitzsimons and Moss who showed that saturating the actin filaments with a strong-binding derivative of myosin, NEM-S1, reduces the length-dependence of Ca^{2+} sensitivity (Fig. 4B).[23] It has been suggested that the strong-binding crossbridge actually increases the affinity of TnC for Ca^{2+} through protein-protein interactions within the thin filament.[24] However, the relationship between cooperative activation and length-dependent activation is still not fully understood.[25,26]

The crossbridge plays a pivotal role in the length-tension relationship of cardiac muscle, and in a quest to understand this, the search has focused on factors that increase the probability of crossbridge formation. Aside from the degree of overlap of actin and myosin within the sarcomere, there are a number of factors that affect the probability of crossbridge formation. One of these, which has received a great deal of attention in recent years, is the degree of lateral spacing between the thick and thin filaments (the lattice spacing). Because myofibrils maintain close to constant volume, an increase in sarcomere length is accompanied by a decrease in cross-sectional area that brings the filaments closer together. It is thought that under these circumstances, the closer proximity of the myosin head to the actin filament increases the probability of strong-binding crossbridge formation.[27] In 1981, Godt and Maughan showed that dextran-induced osmotic compression of the myofilaments of skeletal muscle increased myofilament Ca^{2+} sensitivity.[28] Similar findings have been made in cardiac muscle.[29-31] For example, osmotic compression with 2.5% dextran has been shown to increase myofilament Ca^{2+} sensitivity in skinned cardiac myocytes and muscle,[29,31] and the length-dependent shift in sensitivity is absent when a constant muscle width/lattice spacing is maintained in skinned cardiac muscle by varying the concentration of dextran at different SLs (Fig. 5A).[30]

Although it is clear that manipulation of lattice spacing can modulate the Ca^{2+} sensitivity of the myofilaments, there is some doubt as to whether lattice spacing is the primary mechanism underlying the length-tension relationship in cardiac muscle. Early studies that supported the role of lattice spacing used muscle width as an index of interfilament spacing.[29,30] However, when X-ray diffraction was used to measure lattice spacing directly in skinned cardiac muscle, Konhilas et al observed an increased myofilament Ca^{2+} sensitivity (indexed as both $[Ca^{2+}]_{50}$ and pCa_{50}) when SL was increased from 2.02 to 2.19 μm, but no change in sensitivity when 1% dextran was used to produce the same decrease in lattice spacing (from 43 to 42 nm).[32] The use of X-ray diffraction has revealed that the relationship between lattice spacing and myofilament Ca^{2+} sensitivity is not linear. Konhilas et al and Cazorla et al have shown that when lattice spacing is less than 38 nm or greater than 42-44 nm myofilament Ca^{2+} sensitivity is independent of spacing (see Fig. 5B).[31,32] In the presence of 3% dextran, lattice spacing is reduced beyond that seen as a result of stretch in skinned cardiac muscle (Fig. 5B).[32] Konhilas et al concluded that interfilament spacing is not the factor determining length-dependence of myofilament Ca^{2+} sensitivity in cardiac muscle.[32] There are a few reservations to this bold conclusion. It is well established that lattice spacing is greater in skinned muscle than in intact muscle,[33] and this may cause us to question the physiological relevance of matching lattice spacing with dextran to the effect of stretch in skinned muscle. Secondly, it has been shown (in skeletal muscle) that osmotic pressure has a direct effect on crossbridges and causes collapse of the heads against the myosin backbone.[34] This may counteract the effect of dextran on lattice spacing, and account for the lack of effect of some concentrations of dextran on myofilament sensitivity.

The search for factors that contribute to length-dependent modulation of myofilament Ca^{2+} sensitivity has focused in recent years on the giant sarcomeric protein titin. In terms of mechanotransduction, titin is an important protein. It forms a structural link between the thick myosin filament and the Z-line (see Fig. 1), with 6 titin molecules binding to each half of the thick filament. It also binds to the thin filament at the Z-line[35] and thus has a crucial role in detecting and transmitting forces. The segment of titin within the I band of the sarcomere contains Ig domains and a segment rich in proline, glutamate, valine and lysine residues (the PEVK domain). It is this I-band region that is thought to function as a molecular spring.[36] In

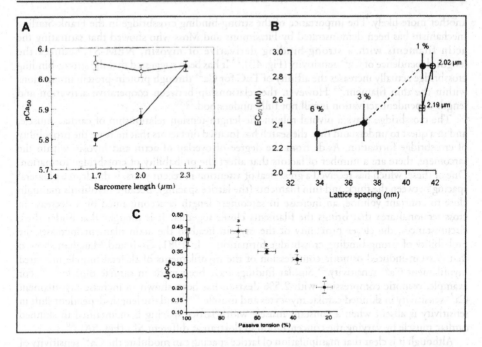

Figure 5. A) The pCa$_{50}$ in skinned bovine trabeculae increases with SL (●); this effect is abolished when muscle width (used as the index of lattice spacing) is kept constant during stretch by varying degrees of osmotic compression with dextran (0) (from Fuchs and Wang).[30] B) In skinned rat trabeculae stretch from a SL of 2.02 (0) to 2.19 μm (□) decreases myofilament lattice spacing and EC$_{50}$ ([Ca^{2+}]$_{50}$). Osmotic compression with 1% dextran (◆) has similar effects on lattice spacing, but the relationship between lattice spacing and EC$_{50}$ is not constant and compression of the lattice space by an equivalent amount to that seen during stretch does not reduce EC$_{50}$. This suggests that lattice spacing is not the primary mechanism underlying the stretch-induced fall in EC$_{50}$ (from Konhilas et al).[32] C) The length-dependent shift in the pCa$_{50}$ of skinned rat ventricular myocytes, following stretch from 1.9 to 2.3 μm (see Fig. 3B), decreases as passive tension is reduced by trypsin digestion of titin (from Cazorla et al).[37]

terms of the Frank-Starling mechanism, Cazorla et al demonstrated that the shift in pCa$_{50}$ in skinned rat ventricular myocytes stretched from a SL of 1.9 to 2.3 μm was decreased in proportion to the decrease in passive tension seen following mild trypsin digestion (Fig. 5C).[37] At SLs that equate to most of the ascending limb of the length-tension relationship, resting tension is largely determined by titin in both cells and muscle strips[38,39] and protein electrophoresis by Cazorla et al[31,37] demonstrated that titin was the main target of their mild trypsin treatment. The positive relationship between passive tension and stretch-induced changes in pCa$_{50}$ was confirmed in skinned mouse myocytes where passive tension was varied either by SL history prior to stretching or trypsin treatment.[31] Titin digestion has also been shown to reduce the length-dependence of myofilament Ca^{2+} sensitivity in skinned rat trabeculae by Fukuda et al.[40]

The evidence for a role of titin in the length-tension relation of cardiac muscle is convincing, but it is not yet clear how titin might exert its effects. It has been argued that titin controls lattice spacing by exerting radial stress that pulls thick and thin filaments closer together as sarcomere length is increased. Indeed, Cazorla et al showed that removal of titin by trypsin digestion in skinned mouse trabeculae increased lattice spacing by around 3 nm throughout the range of SLs on the ascending limb of the length-tension curve (1.9-2.4 μm).[31] Fukuda et al showed that in skinned rat trabeculae an increase in SL from 1.9 to 2.3 μm caused a 12% decrease in muscle diameter and a leftwards shift in pCa$_{50}$, and that these effects could be mimicked by 6% dextran suggesting an important role for lattice spacing, but were insensitive to trypsin, consistent with the lack of involvement of titin.[40] They concluded that in preparations where there is a low collagen content (such as a skinned cardiac myocyte), titin may work through changing lattice spacing. At long SLs, collagen is increasingly important in passive tension generation,[38] and increased collagen content of preparations such as trabeculae might resist the lattice expanding effect of titin digestion. However it should be remembered that the work of Konhilas et al brings into question the extrapolation of changes in muscle width to equivalent changes in lattice spacing, and the role of lattice spacing in the length-dependent modulation of Ca^{2+} sensitivity.[32]

Given the convincing evidence for the importance of titin in the Frank-Starling mechanism, how else might titin work if not through modulation of interfilament spacing? Titin may also regulate thick filament structure and myosin cross-bridge arrangement. It has been suggested that titin strain increases the disorder of myosin heads within the thick filament, which in turn increases the probability of actin and myosin interacting.[26,31,40]

Inhomogeneities in Length-Tension Relationships

It is well established that diastolic and systolic SL are not uniform across the ventricular wall (e.g., ref. 41), thus in vivo tension during the cardiac cycle in different regions of the heart will vary depending on changes in SL, external loading and the intrinsic length-tension relationships.[42] It is also interesting to note that ventricular dilation increases diastolic length but can reduce end-systolic length due to the increase in contractility.[43] Thus following ventricular dilation there will be a greater dynamic change in diastolic to systolic myofilament Ca^{2+} sensitivity. Differences in the shape of the length-tension relationship can occur within isolated regions of the ventricle or across the whole ventricle in response to physiological and pathological interventions (such as systolic lengthening of border zone muscle of ventricular aneurysms).[44] In myocytes isolated from the left ventricle of exercised rats, Natali et al reported a steepening of the length-tension relationship in intact sub-endocardial cells.[45] Diffee and Nagle showed an increased stretch-induced shift in the pCa_{50} in skinned myocytes from exercise-trained animals compared with sedentary controls (Fig. 6A), that was greater in sub-endocardial cells.[46,47]

Titin may play a role in the regional variation of some mechanical properties of cardiac myocytes. The mammalian myocardium coexpresses two distinct titin isoforms, a smaller stiffer isoform which contains a region known as N2B (N2B titin is expressed exclusively in the heart), and a larger more supple titin isoform containing both N2B and N2A unique sequences (N2A titin is also expressed in skeletal muscle). Cazorla et al demonstrated that the relative expression of the supple N2BA isoform increases with species in line with the size of the heart, and that there may also be variations in expression ratios of N2BA and N2B across the ventricular wall.[48] It is intriguing for mechanical function that coexpression of titin isoforms occurs within the same myocyte,[48] and even within the same half-sarcomere.[49] Isoform expression would be predicted to influence myocardial stiffness, and thereby the rate of ventricular filling and end-diastolic volume for a given filling pressure.[48]

The Length-Tension Relationship in Human Cardiac Disease

The Frank-Starling mechanism provides a means for the heart to increase its contractile performance. How then is this fundamental process affected in cardiac disease? The literature regarding the effect of human heart failure on the Frank-Starling response is somewhat diverse. In 1994, Schwinger et al reported that the length-tension relationship was absent in skinned muscle preparations from hearts of patients with dilated cardiomyopathy.[50] Others have observed that there is a length-tension relationship in tissue from the failing human heart, but that it is more shallow.[51,52] Morano et al showed that heart failure was associated with a down-regulation of titin expression which could account for the attenuation of length-dependent force development, given the evidence for a role of titin in this.[51] By contrast, in the intact failing heart ventricular performance has been shown to change with preload and the full effectiveness of the Frank-Starling mechanism has been demonstrated (Fig. 6B).[53] Clearly experiments using human tissue are limited by the supply of tissue and the availability of nonfailing tissue for comparison.

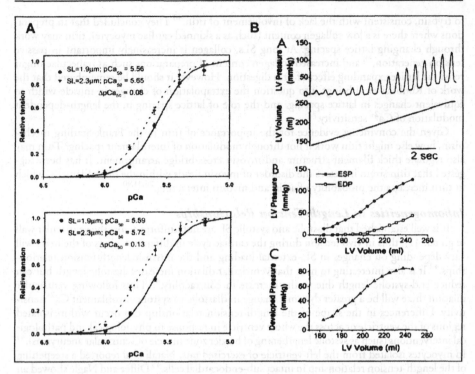

Figure 6. The effect of physiological and pathological interventions on the length-tension relationship. A) The length-dependent shift in pCa_{50} is greater in skinned ventricular myocytes from rats undertaking chronic aerobic exercise (lower panel) than that seen in myocytes from sedentary animals (upper panel) (from Diffee and Nagle).[46] B) The relationship between ventricular pressure and volume in an intact heart from a patient with dilated cardiomyopathy. The Frank-Starling mechanism is preserved in the failing heart as evidenced by increased left ventricular (LV) pressure with ventricular volume. Developed pressure is end-systolic pressure (ESP) minus end-diastolic pressure (EDP) (from Holubarsch et al).[53]

Furthermore there is evidence that several factors, including tissue damage during isolation, lack of sarcomere length control and the influence of changes in the extracellular matrix in failing tissue, may contribute to the diversity of the literature in this area.[53,54]

Stretch and the Slow Increase in Active Force

Upon stretch of the myocardium the immediate increase in force is followed by a slower increase in force that takes several minutes to stabilise. In isolated ventricular muscle, this slow increase in force is accompanied by a corresponding slow increase in the amplitude of the $[Ca^{2+}]_i$ transient,[9] which has also been reported in single ventricular myocytes,[55] atrial muscle,[56] and the intact heart (see Fig. 7).[57] In ventricular muscle and myocytes, the mechanisms associated with this response appear to differ with species and with preparation (see refs. 58 and 59). However it is clear that, by contrast to the immediate increase in force upon stretch which is sarcomere-based, the mechanism underlying the slow response is membrane-dependent. Signalling molecules that have been implicated include cyclic AMP (cAMP),[60,61] angiotensin II,[62] endothelin 1,[62,63] and nitric oxide.[64] It is still unclear how cAMP may contribute to the slow increase in $[Ca^{2+}]_i$, as the slow response is independent of two of the primary targets of cAMP-dependent modulation of $[Ca^{2+}]_i$, $I_{Ca,L}$ (although this has never been shown directly) and phospholamban.[60,65] In intact ventricular muscle activation of the NaH antiport, thereby stimulating reverse mode NaCa exchange and increasing $[Ca^{2+}]i$, has been shown to contribute

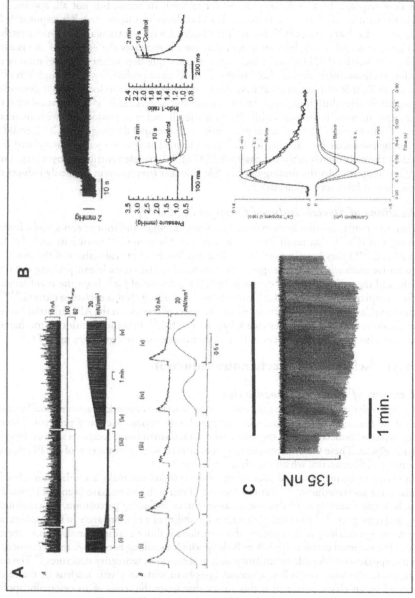

Figure 7. The slow increase in contractility and [Ca²⁺]ᵢ following stretch in various cardiac preparations. A) [Ca²⁺]ᵢ (measured using aequorin) and force in a cat papillary muscle held at 82% and 100% L_max. Upper traces show slow timebase recordings, lower traces show averages from selected time points indicated above. The slow change in the [Ca²⁺]ᵢ transient is revealed by comparison of (iv) and (v) (from Allen and Kurihara).[9] B) Temporal changes in developed pressure and [Ca²⁺]ᵢ (measured using Indo-1) recorded in the isolated rat atrium in response to an increase in intra-atrial pressure (from Tavi et al).[56] C) Active force and [Ca²⁺]ᵢ (measured with fura-2) recorded in a single rat ventricular myocyte in response to stretch using a pair of carbon fibres. The slow increase in contractility is associated with a slow increase in the amplitude of the [Ca²⁺]ᵢ transient (from Hongo et al).[55]

to the slow response to stretch, (see ref. 58 for review). In some, but not all, species, it is thought the sequential release of angiotensin II and endothelin 1 upon stretch is responsible for activation of the NaH antiport.[58] From data obtained with ventricular myocytes stretched within an agarose gel, it has been proposed that stretch increases the activity of the enzyme phosphatidyl inositol-3-OH kinase which causes nitric oxide-dependent s-nitrosylation of the ryanodine receptor, increasing the Ca^{2+} release channel open probability,[64] although it is difficult to reconcile this with studies that have shown the slow increase in force to be independent of sarcoplasmic reticulum function.[12] In atrial tissue the increase in $[Ca^{2+}]_i$ associated with the slow increase in force is present within 10s of a stretch and is not associated with increased cAMP.[56,66] Interestingly, in these atrial preparations, the temporal changes in $[Ca^{2+}]_i$ could be modelled by enhanced TnC affinity for Ca^{2+} (seen immediately upon stretch) associated with Ca^{2+} entry through stretch-activated channels (SACs).[56] In single ventricular myocytes, a role for SACs is supported by the finding that the SAC blocker streptomycin markedly reduces the slow increase in force seen upon stretch.[67]

Implications of Stretch-Induced Changes in $[Ca^{2+}]_i$

When ventricular dilation leads to increased diastolic length and shorter end-systolic length, the changes in $[Ca^{2+}]_i$ that result (from enhanced myofilament Ca^{2+} sensitivity and the slow increase in $[Ca^{2+}]_i$) may have important implications for the electrical activity of the heart, via Ca^{2+} sensitive channels and exchangers. Shortening of cardiac muscle can prolong the Ca^{2+} transient and the action potential duration (APD) via release of Ca^{2+} from the myofilaments into the sarcoplasm and increased Ca^{2+} extruding, depolarising, NaCa exchange current.[68,69] A reduction in $[Ca^{2+}]_i$ has been shown to reduce the effect of stretch on the APD in atrial tissue[70] and in single ventricular myocytes (see Chapter by Lab).[71] Inhomogeneities in mechanical activity may also provoke differences in $[Ca^{2+}]_i$ that lead to inhomogeneities in APD.[72]

The Cytoskeleton and Mechanotransduction

Mechanisms of Mechanotransduction

The cytoskeleton of the cardiac cell consists of an intricate network of actin microfilaments, intermediate filaments and microtubules associated with various accessory proteins. The cytoskeleton forms connections with the cell exterior via transmembrane receptors such as integrins and dystroglycan. These transmembrane receptors interact with components of the ECM, such as laminin and fibronectin, which sheath the myocyte.

The structure, connections and associations of the cytoskeleton make it an ideal candidate for a cellular mechanotransducer.[73,74] Various theories of cell architecture have been put forward to explain how cells transmit and balance mechanical forces, including percolation, the continuum model, and tensegrity.[75-78] Percolation describes the ability of a random array of interconnecting structures to signal as long as the number of connections is above a critical threshold (i.e., there is a physically connected pathway from A to B along which signalling between A and B can occur). It can incorporate considerable redundancy and, in some cases, tensegrity structures.[75] The continuum model describes the cell as a viscous cytoplasm with an elastic nucleus in the centre surrounded by an elastic cortex; it is this cortex that bears the stress of an externally applied mechanical stimulus.[79] Tensegrity (tensional integrity) is based principally upon a structural arrangement that has elements that resist compression connected to tension producing elements that generate prestress. Ingber has put forward a case for the tensegrity model being most consistent with experimental data.[77,78] In a generic cell, the compression elements may be microtubules, thick actin bundles, and the external connections of the cell to the ECM, whilst the tension producing elements are actin microfilaments and intermediate filaments (Fig. 8). The cortical cytoskeleton is thought to be a separate tensegrity structure. In the cardiac myocyte, we must add to this scheme the actomyosin cross bridges that generate tension, and the Z-line structures and titin which act as a resistance to compression when the myocyte is shortened beyond slack length but also generate resistance to stretching. Consistent with one of the tenets of tensegrity (but not

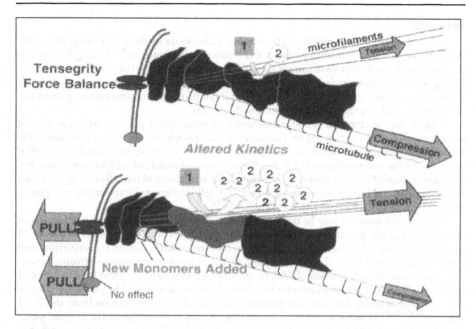

Figure 8. Possible mechanism of mechanotransduction conveyed via the principle of tensegrity in the cytoskeleton. Upper panel) counterbalanced force in tensed microfilaments and compressed microtubules. ECM receptors which span the sarcolemma (double line), such as integrins (black ovals) link the ECM and regulatory proteins (black structures); some ECM receptors are not linked to the cytoskeleton (gray oval). Lower panel) application of force to receptor linked to the cytoskeleton is transmitted to the cytoskeleton where extra sub-units may be added to decompressed microtubules. Signalling proteins are physically distorted and, as a result, the conversion of substrate (1) into product (2) is increased (from Ingber).[78]

to the exclusion of other models[76]), the whole heart is prestressed as when a ventricle is cut it springs open (e.g., ref. 41), and isolated myocytes generate resting tension (e.g., ref. 38).

Understanding how cells transmit and balance mechanical force provides insight into the way that cells can transduce mechanical force into biological activity. Tensegrity theory can be used to explain mechanosensitivity conveyed via the cytoskeleton by the linking of modular tensegrity structures (see refs. 77,78,80,81). Changes in the compressive or tension producing elements are thought to influence the signalling molecules that are bound to cytoskeletal proteins; thus a mechanical stimulus is transduced into a chemical signal.[74] Such mechanisms may influence the signalling pathways discussed in other chapters of this volume. We will now consider the evidence that the ECM-cytoskeleton axis is involved in mechanotransduction in the myocardium.

Extracellular Matrix

The cardiac myocyte is sheathed in an ECM that is made up of proteins such as collagen, fibronectin, and laminin that are secreted primarily by fibroblasts (see refs. 82-84 for reviews). The ECM is a dynamic entity that plays a vital role in the structure and function of the myocardium.[82] The ECM forms part of a pathway that links the exterior of the cell with the interior, allowing bi-directional transmission of information. The connection between the ECM and the cell interior is provided by transmembrane ECM receptors such as integrins and dystroglycan. In response to pressure overload, the expression ratios of various ECM proteins are altered and this has consequences for compliance of the ventricles (e.g., ref. 85). The importance of the pathway between ECM and cell interior is underscored by the finding that this linkage is disturbed in patients with dilated cardiomyopathy (see ref. 86 for a review).

Integrins

Integrins are heterodimers of α and β integrins, with specificity for matrix components and signalling properties determined by the α and β subtype combination.[87-89] Integrins have no enzymatic activity of their own; instead their short cytoplasmic tails transduce signals by associating with other proteins including actin binding proteins (e.g., vinculin, talin, α-actinin), nonreceptor protein tyrosine kinases (e.g., focal adhesion kinase), transmembrane growth factor receptors, and G proteins (see reviews refs. 83,86,90).

Integrins are concentrated in costameres which are specialised regions of the sarcolemma at or near the Z-line.[83] Costameres are composite structures which have a dual role within the terminally differentiated cardiac cell. Components which act to tether superficial myofibrils at the Z-line to the cell membrane (such as α-actinin, desmin intermediate filaments) allow lateral transmission of force developed by the contractile machinery of the cell (see Fig. 1). However, this structural link between the ECM and the cell interior at the costamere also plays a role in the transduction of mechanical stimuli.[83,91,92]

A mechanical stimulus may be transduced into a chemical signal via integrin-activation of intracellular signalling pathways. The association of integrins with components of the ECM causes clustering of integrins in the cell membrane, which results in activation and autophosphorylation of focal adhesion kinase (FAK), a nonreceptor protein tyrosine kinase. The continuing phosphorylation of FAK creates docking sites to which other kinases can bind (see Fig. 1).[93] Many signalling pathways have been shown to be activated by the ECM-integrin axis: Tyrosine kinases; Ser and Thr kinases; phosphatidyl inositol metabolism; the NaH antiport.[86,87,94,95] Directly relevant to the sensing of stretch in the heart, the enzyme phosphatidyl inositol-3-OH kinase, which has been implicated in the slow response to stretch, has been shown to associate with FAK only minutes after inducing pressure overload in the intact rat heart.[96] The integrin pathway has also been implicated in the activation of signalling pathways which lead to the development of cardiac hypertrophy in response to mechanical stretch.[90]

Mechanotransduction may also occur via ECM-integrin links with the cytoskeleton. Autophosphorylation of FAK may promote association of cytoskeletal elements with the focal adhesion complex.[93] Pulling on surface integrin receptors has been shown to trigger cytoskeletal organisation in endothelial cells.[97] The actin cytoskeleton has been implicated in the integrin-dependent modulation of signalling pathways (e.g., the β-adrenergic pathway).[98] As well as allowing the cell to respond to mechanical stimuli, the integrins themselves are sensitive to the forces that they transmit; in the cardiac cell, force regulates cellular levels of integrins, and possibly their association with focal adhesions.[99]

Dystroglycan

There are other proteins which act as ECM-receptors in the cardiac cell. One of these is the dystroglycan complex, consisting of β-dystroglycan, an integral membrane glycoprotein, which associates with the sarcoglycan complex within the cell membrane. β-dystroglycan interacts with α-dystroglycan on the external surface of the cell which in turn binds to the ECM protein laminin.[100] On the cytoplasmic face of the membrane, β-dystroglycan tethers the actin-binding protein dystrophin (see refs. 101,102 for reviews). Dystrophin may also link the intermediate filament protein desmin to the plasma membrane.[103] In the human myocyte, dystrophin, like integrin, is concentrated in costameres.[104] The importance of the dystrophin link between the ECM and cytoskeleton is underscored by the observation of cardiac (and skeletal) muscle degeneration in both Duchenne and Becker's muscular dystrophy in which the gene encoding dystrophin is mutated.[105] It has been proposed that dystrophin binds to a population of cytoskeletal γ-actin which is present as a sub-membranous cortical lattice,[101] as well as to sarcomeric α-actin of the thin filament.[103] Evidence suggests that the transmembrane linkage with cortical actin is important in stabilisation of the surface sarcolemma, and that it is the absence of such stability which leads to muscle damage and degeneration seen in muscular dystrophy.[102] However, through its costameric connections with the Z-line (either via actin myofilaments or desmin), dystrophin has also been ascribed a role in lateral force transmission.[106]

Actin

Visualisation of the actin cytoskeleton in adult cardiac myocytes is difficult,[103] however, it is thought that cytoskeletal actin exists as a sub-membranous cortical lattice, which has been well-described in other cells such as the erythrocyte,[107] or as actin microfilament bundles which are prominent in cultured neonatal cardiac myocytes.[108] Of the 3 main cytoskeletal components it is primarily the actin microfilaments which appear to form an important intracellular connection with integrins, although microtubules and intermediate filaments may also be involved.[97,109,110] Many of the proteins with which integrins associate at the costamere are actin-binding proteins (e.g., vinculin, talin, α-actinin).[83,86] Actin forms an important component of the focal adhesion in cultured cells, although it has not been detected as part of the ECM-integrin complex in vivo.[83] Sparse cortical actin has been labelled in the adult cardiac myocyte.[111] Besides actin, the primary constituents of this cortical lattice are the actin binding proteins spectrin and dystrophin. In skeletal muscle, cortical γ-actin filaments have been observed,[112,113] which are absent in the dystrophin-deficient *mdx* mouse.[113] Tensegrity theory implies that interactions with the cortical actin network are unlikely to result in transmission of mechanical signals deep into the cell interior because stress transmitted to an elastic element will dissipate locally.[81]

Cytoskeletal actin, in turn, forms attachments with regulatory proteins within the cell and with ion channels and exchangers (see refs. 114,115 for reviews). There is evidence that the mechanical signal that activates SACs (see Baumgarten chapter) may be modulated by the actin cytoskeleton. Cytochalasin B and cytochalasin D disrupt the actin cytoskeleton and have been shown to increase the sensitivity of SACs to stretch (Fig. 9A).[116,117] This is thought to be brought about by a reduction in the resistance of the membrane to distension due to the breakage of tethering cytoskeletal elements.[74]

Intermediate Filaments

Desmin is the major protein component of cardiac intermediate filaments. They are formed of a dimer composed of two α-helical chains orientated in parallel and intertwined in a coiled-coil rod, around 10 nm in diameter.[118] Intermediate filaments provide a physical link between the Z-line of peripheral sarcomeres and the sarcolemma at costameres (and via these to the ECM), between Z-lines of adjacent sarcomeres, between the Z-line and the nucleus, and in cell-to-cell contacts at the desmosome (see Fig. 1).[110] Transgenic desmin-deficient mice hearts are hypertrophied, and have decreased systolic function,[119,120] and it is suggested that desmin maintains both series (within the myofibril) and parallel (between myofibrils) sarcomere alignment, and that the absence of desmin at the Z-line of the sarcomere impairs the generation or transmission of force through misalignment of sarcomere units. The physical pathway between sarcomere and nuclear membrane which desmin provides could enable changes in sarcomere length to be translated into changes in gene expression. Bloom et al propose that this mechanism may work through stretch-induced changes in chromatin associated with the nuclear envelope.[121]

Microtubules

Physical interactions between microtubules and sarcomeres under normal conditions are not well understood. Microtubules do make a small contribution to the passive properties of cardiac myocytes.[38] In a linear polymer, such as a microtubule, mechanical stimulation such as an extending force is predicted to lower the critical concentration for sub-unit (αβ tubulin heterodimer) assembly and enhance polymer (microtubule) stability.[122,123] This property may be important in certain pathological conditions, such as pressure overload, that induce hypertrophy of the myocardium. It seems that when hypertrophy can no longer compensate for the increase in pressure, there is an increase in wall stress and that, under these conditions, microtubules proliferate (see Cooper[124] for a review). This proliferation of microtubules is associated with an increase in the internal apparent viscosity of myocytes,[125,126] and a decrease in contractility that can be reversed by microtubule disruption (Fig. 9B).[124,127,128] The decrease in contractility is mimicked by chemical proliferation of microtubules.[128,129] These findings have been explained

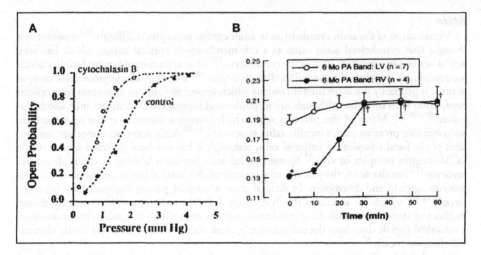

Figure 9. A) Pressure increases the probability of opening in rat atrial stretch-activated channels; the sensitivity to pressure is increased by treatment with the actin filament disruptor cytochalasin B (10 μM, for 6h) (from Kim).[117] B) Pulmonary artery (PA) banding in cats induces pressure overload which results in hypertrophy, microtubule proliferation and reduced myocyte contractility at 6 months in the right ventricle (RV) but not in the left ventricle (LV). Treatment with the microtubule disruptor colchicine does not modify contractility in LV myocytes but restores depressed contractility in RV myocytes (from Tsutsui et al).[128]

by increased resistance to sarcomere shortening from the proliferated microtubules (see ref. 124). This effect upon microtubules is not seen in all models of hypertrophy, but does appear relevant to certain situations of human pressure-overload hypertrophy and failure.[130-133]

Conclusion

In this chapter we have dealt with the role of sarcomeric and cytoskeletal proteins in the response of cardiac muscle to a mechanical stimulus such as stretch. When sarcomere length is increased, myofilament Ca^{2+} sensitivity is enhanced, leading to an increase in contractile force. Ultimately it appears to be an increase in the probability of actomyosin crossbridge formation that underlies this length-dependent activation. Increased myosin head disorder as a result of titin strain, and positive cooperativity of crossbridge binding are likely candidates for this mechanism; the role of interfilament spacing is more controversial. Although the processes responsible for the immediate response to stretch reside entirely within the sarcomere, the slow secondary increase in contractility appears to be primarily a membrane-based response. Multiple signalling pathways have been implicated at the level of the sarcolemma (NaH antiport activation, Ca^{2+} entry through stretch-activated channels), and the sarcoplasmic reticulum (s-nitrosylation of the ryanodine receptor).

The connections that exist between the extracellular matrix, and the cell interior (via both integrins and β-dystroglycan) form a vital pathway that allows the cell to sense and respond to mechanical stimuli such as stretch. Mechanical signals may be transmitted via elements of the cytoskeleton, or through activation of signalling cascades at the focal adhesion site. Connections between the contractile machinery of the cell and the sarcolemma and extracellular matrix also allow the lateral and longitudinal transmission of force developed by the sarcomere. The importance of the ECM-cytoskeleton axis is clearly illustrated by the profound impact that diseases such as muscular dystrophy have on normal cardiac structure and function.

Acknowledgement
This work was supported by The British Heart Foundation.

References

1. Frank O. Zur dynamik des herzmuscckels. Z Biol 1895; 32:370-447.
2. Patterson S, Starling EH. On the mechanical factors which determine the output of the ventricles. J Physiol 1914; 48:357-337.
3. Bainbridge FA. The influence of venous filling upon the rate of the heart. J Physiol 1915; 50:65-84.
4. Gordon AM, Huxley AF, Julian FJ. The variation in isometric tension with sarcomere length in vertebrate muscle fibres. J Physiol 1966; 184:170-192.
5. Allen DG, Jewell BR, Murray JW. The contribution of activation processes to the length-tension relation of cardiac muscle. Nature 1974; 248:606-607.
6. Fabiato A, Fabiato F. Dependence of the contractile activation of skinned cardiac cells on the sarcomere length. Nature 1975; 256:54-56.
7. Allen DG, Kentish JC. The cellular basis of the length-tension relation in cardiac muscle. J Mol Cell Cardiol 1985; 17:821-840.
8. Lakatta EG. Length modulation of muscle performance: Frank-Starling law of the heart. In: Fozzard HA, Haber E, Jennings RB et al, eds. The Heart and Cardiovascular System, 2nd ed. New York: Raven Press,1992:1325-1352.
9. Allen DG, Kurihara S. The effects of muscle length on intracellular calcium transients in mammalian cardiac muscle. J Physiol 1982; 327:79-94.
10. Calaghan SC, White E. The role of calcium in the response of cardiac muscle to stretch. Prog Biophys Mol Biol 1999; 71:59-90.
11. Hibberd MC, Jewell BR. Calcium and length-dependent force production in rat ventricular muscle. J Physiol 1982; 329:527-540.
12. Kentish JC, Wrzosek A. Changes in force and cytosolic Ca^{2+} concentration after length changes in isolated rat ventricular trabeculae. J Physiol 1998; 506:431-444.
13. Hofmann PA, Fuchs F. Effect of length and cross-bridge attachment on Ca^{2+} binding to cardiac troponin C. Am J Physiol 1987; 253:C90-C96.
14. Babu A, Sonnenblick E, Gulati J. Molecular basis for the influence of muscle length on myocardial performance. Science 1988; 240:74-76.
15. Gulati J. Is cardiac troponin C the length sensor in Starling's law of the heart. Trends Cardiovasc Med 1996; 6:245-246.
16. Akella AB, Su H, Sonnenblick EH et al. The cardiac troponin C isoform and the length dependence of Ca^{2+} sensitivity of tension in myocardium. J Mol Cell Cardiol 1997; 29:381-389.
17. Moss RL, Nwoye LO, Greaser ML. Substitution of cardiac troponin C into rabbit muscle does not alter the length-dependence of Ca^{2+} sensitivity of tension. J Physiol 1991; 440:273-289.
18. McDonald KS, Field LJ, Parmacek MS et al. Length-dependence of Ca^{2+} sensitivity of tension in mouse cardiac myocytes expressing skeletal troponin C. J Physiol 1995; 483:131-139.
19. Kajiwara H, Morimoto S, Fukuda N et al. Effect of troponin I phosphorylation by protein kinase A on length-dependence of tension activation in skinned cardiac muscle fibers. Biochem Biophys Res Comm 2000; 272:104-110.
20. Allen DG, Kentish JC. Calcium concentration in the myoplasm of skinned ferret ventricular muscle following changes in muscle length. J Physiol 1988; 407:489-503.
21. Lehrer SS, Geeves MA. The muscle thin filament as a classical cooperative/allosteric regulatory system. J Mol Biol 1998; 277:1081-1089.
22. Bremel RD, Weber A. Cooperation within actin filament in vertebrate skeletal muscle. Nat New Biol 1972; 238:97-101.
23. Fitzsimons DP, Moss RL. Strong binding of myosin modulates length-dependent Ca^{2+} activation of rat ventricular myocytes. Circ Res 1998; 83:602-607.
24. Fuchs F, Smith SH. Calcium, crossbridges and the Frank Starling relationship. News Physiol Sci 2001; 16:5-10.
25. Konhilas JP, Irving TC, de Tombe PP. Frank-Starling law of the heart and the cellular mechanisms of length-dependent activation. Pflugers Archiv 2002; 445:305-310.
26. Moss RL, Fitzsimons DP. Frank-Starling relationship: Long on importance, short on mechanism. Circ Res 2002; 90:111-113.
27. Fukuda N, Kajiwara H, Ishiwata S et al. Effects of MgADP on length dependence of tension generation in skinned rat cardiac muscle. Circ Res 2000; 86:E1-6.
28. Godt RE, Maughan DW. Influence of osmotic compression on calcium activation and tension in skinned muscle fibers of the rabbit. Pflugers Arch 1981; 391:334-337.
29. McDonald KS, Moss RL. Osmotic compression of single cardiac myocytes eliminates the reduction in Ca^{2+} sensitivity of tension at short length. Circ Res 1995; 77:115-119.
30. Fuchs F, Wang Y-P. Sarcomere length versus interfilament spacing as determinants of cardiac myofilament Ca^{2+} sensitivity and Ca^{2+} binding. J Mol Cell Cardiol 1996; 28:1375-1383.

31. Cazorla O, Wu Y, Irving TC et al. Titin-based modulation of calcium sensitivity of active tension in mouse skinned cardiac myocytes. Circ Res 2001; 88:1028-1035.
32. Konhilas JP, Irving TC, de Tombe PP. Myofilament calcium sensitivity in skinned rat cardiac trabeculae: Role of interfilament spacing. Circ Res 2002; 90:59-65.
33. Irving TC, Konhilas J, Perry D et al. Myofilament lattice spacing as a function of sarcomere length in isolated rat myocardium. Am J Physiol 2000; 279:H2568-H2573.
34. Irving TC, Millman BM. Changes in thick filament structure during compression of the filament lattice in relaxed frog sartorius muscle. J Muscle Res Cell Motil 1989; 10:385-94.
35. Trombitas K, Granzier H. Actin removal from cardiac myocytes shows that near Z line titin attaches to actin while under tension. Am J Physiol 1997; 273:C662-70.
36. Helmes M, Trombitas K, Granzier H. Titin develops restoring force in rat cardiac myocytes. Circ Res 1996; 79:619-26.
37. Cazorla O, Vassort G, Garnier D et al. Length modulation of active force in rat cardiac myocytes: Is titin the sensor? J Mol Cell Cardiol 1999; 31:1215-1227.
38. Granzier HL, Irving TC. Passive tension in cardiac muscle: Contribution of collagen, titin, microtubules and intermediate filaments. Biophys J 1995; 68:1027-1044.
39. Wu Y, Cazorla O, Labeit D et al. Changes in titin and collagen underlie diastolic stiffness diversity of cardiac muscle. J Mol Cell Cardiol 2000; 32:2151-2162.
40. Fukuda N, Sasaki D, Ishiwata S et al. Length dependence of tension generation in rat skinned cardiac muscle: Role of titin in the Frank-Starling mechanism of the heart. Circulation 2001; 104:1639-1645.
41. Rodriguez EK, Omens JH, Waldman, LK et al. Effect of residual stress on transmural sarcomere length distributions in rat left ventricle. Am J Physiol 1993; 264:H1048-1056.
42. Solovyova O, Katsnelson L, Guriev S et al. Mechanical inhomogeniety of myocardium studied in parallel and series cardiac muscle duplexes: Experiments and models. Chaos Solitons Fractal 2002; 13:1685-1711.
43. Lab M. Mechanosensitivity as an integrative system in the heart: An audit. Prog Biophys Mol Biol 1999; 71:7-27.
44. Moulton MJ, Downing SW, Creswell LL et al. Mechanical dysfunction in the border zone of an ovine model of left ventricular aneurysm. Ann Thorac Surg 1995; 60:986-998.
45. Natali AJ, Wilson LA, Peckham M et al. Different regional effects of voluntary exercise on the mechanical and electrical properties of rat ventricular myocytes. J Physiol 2002; 541:863-875.
46. Diffee GM, Nagle DF. Exercise training alters length dependence of contractile properties in rat myocardium. J Appl Physiol 2003; 94:1137-1144.
47. Diffee GM, Nagle D. Regional differences in effects of exercise training on contractile and biochemical properties of rat cardiac myocytes. J Appl Physiol 2003; 95:35-42.
48. Cazorla O, Freiburg A, Helmes M et al. Differential expression of cardiac titin isoforms and modulation of cellular stiffness. Circ Res 2000; 86:59-67.
49. Trombitas K, Wu Y, Labeit D et al. Cardiac titin isoforms are coexpressed in the half-sarcomere and extend independently. Am J Physiol 2001; 281:H1793-H1799.
50. Schwinger RH, Bohm M, Koch A et al. The failing human heart is unable to use the Frank-Starling mechanism. Circ Res 1994; 74:959-969.
51. Morano I, Hadicke K, Grom S et al. Titin, myosin light chains and C-protein in the developing and failing human heart. J Mol Cell Cardiol 1994; 26:361-368.
52. Vahl CF, Timek T, Bonz A et al. Length dependence of calcium and force transients in normal and failing human myocardium. J Mol Cell Cardiol 1998; 30:957-966.
53. Holubarsch C, Ruf T, Goldstein DJ et al. Existence of the Frank-Starling mechanism in the failing human heart. Investigations on the organ, tissue, and sarcomere levels. Circulation 1996; 94:683-689.
54. de Tombe PP. Altered contractile function in heart failure. Cardiovasc Res 1998; 37:367-380.
55. Hongo K, White E, Le Guennec J-Y et al. Changes in $[Ca^{2+}]_i$, $[Na^+]_i$ and Ca^{2+} current in isolated rat ventricular myocytes following an increase in cell length. J Physiol 1996; 491:599-609.
56. Tavi P, Han C, Weckström M. Mechanisms of stretch-induced changes in $[Ca^{2+}]_i$ in rat atrial myocytes: Role of increase troponin C affinity and stretch-activated ion channels. Circ Res 1998; 83:1165-1177.
57. Tucci PFJ, Bregagnollo EA, Spadaro J et al. Length dependence of activation studied in the isovolumic blood-perfused dog heart. Circ Res 1984; 55:171-188.
58. Cingolani HE, Perez NG, Pieske B et al. Stretch-elicited Na^+/H^+ exchanger activation: The autocrine/paracrine loop and its mechanical counterpart. Cardiovasc Res 2003; 57:953-960.
59. Calaghan SC, Belus A, White E. Do stretch-induced changes in intracellular calcium modify the electrical activity of cardiac muscle? Prog Biophys Mol Biol 2003; 82:82-95.

60. Calaghan SC, Colyer J, White E. Cyclic AMP but not phosphorylation of phospholamban contributes to the slow inotropic response to stretch in ferret papillary muscle. Pflugers Arch 1999; 437:780-782.
61. Todaka K, Ogino K, Gu A et al. Effect of ventricular stretch on contractile strength, calcium transient, and cAMP in intact canine heart. Am J Physiol 1998; 274:H990-H1000.
62. Perez NG, de Hurtado MC, Cingolani HE. Reverse mode of the Na^+-Ca^{2+} exchange after myocardial stretch: Underlying mechanism of the slow force response. Circ Res 2001; 88:376-382.
63. Calaghan SC, White E. The involvement of angiotensin II, endothelin 1 and the endothelium on the slow inotropic effect of stretch in ferret papillary muscle. Pflugers Archiv 2001; 441:514-520.
64. Vila-Petroff MG, Kim SH, Pepe S et al. Endogenous nitric oxide mechanisms mediate the stretch dependence of Ca^{2+} release in cardiomyocytes. Nat Cell Biol 2001; 3:867-873.
65. Chuck L, Parmley CC. Caffeine reversal of the length-dependent changes in myocardial state in the cat. Circ Res 1980; 47:592-598.
66. Tavi P, Weckström M, Ruskoaho H. cAMP- and cGMP-independent stretch-induced changes in the contraction of rat atrium. Pflugers Arch 2000; 441:65-68.
67. Calaghan SC, White E. Signalling pathways which underlie the slow inotropic response to myocardial stretch. J Physiol 2002; 544P:23S.
68. Lab M, Allen D, Orchard CH. The effect of shortening on myoplasmic calcium concentration and on action potential in mammalian ventricular muscle. Circ Res 1984; 55:825-829.
69. Janvier NC, Boyett MR. The role of Na-Ca exchange current in the cardiac action potential. Cardiovasc Res 1996; 32:69-84.
70. Han C, Tavi P, Weckström M. Role of the sarcoplasmic reticulum in the modulation of rat cardiac action potential by stretch. Acta Physiol Scand 1999; 167:111-117.
71. Belus A, White E. Streptomycin and intracellular calcium modulate the response of single guinea pig ventricular myocytes to axial stretch. J Physiol 2003; 546:501-509.
72. Markhasin VS, Solovyova O, Katsnelson L et al. Mechano-electric interactions in heterogeneous myocardium: Development of fundamental experimental and theoretical models. Prog Biophys Mol Biol 2003; 82:207-220.
73. Watson PA. Function follows form: Generation of intracellular signals by cell deformation. FASEB J 1991; 5:2013-2019.
74. Janmey PA. The cytoskeleton and cell signaling: Component localization and mechanical coupling. Physiol Rev 1998; 78:763-81.
75. Forgacs G. On the possible role of cytoskeletal filamentous networks in intracellular signaling: An approach based on percolation. J Cell Sci 1995; 108:2131-2143.
76. Heidemann SR, Lamoureaux P, Buxbaum RE. Opposing views on tensegrity as a structural framework for understanding cell mechanics. J Appl Physiol 2000; 89:1663-1678.
77. Ingber DE. Tensegrity I. Cell structure and hierarchical systems biology. J Cell Sci 2003; 116:1157-1173.
78. Ingber DE. Tensegrity II. How structural networks influence cellular information processing networks. J Cell Sci 2003; 116:1397-1408.
79. Schmidt-Schonbein GW, Kosawada T, Skalak T et al. Membrane model of endothelial cell leukocytes. A proposal for the origin of cortical stress. J Biomechan Eng 1995; 117:171-178.
80. Ingber DE. Tensegrity: The architectural basis of cellular mechanotransduction. Ann Rev Physiol 1997; 59:575-99.
81. Ingber DE. Opposing views on tensegrity as a structural framework for understanding cell mechanics. J Appl Physiol 2000; 89:1663-1678.
82. Pelouch V, Dixon IMC, Golfman L et al. Role of extracellular matrix proteins in heart function. Mol Cell Biochem 1994; 129:101-120.
83. Borg TK, Goldsmith EC, Price R et al. Specialization at the Z line of cardiac myocytes. Cardiovasc Res 2000; 46:277-285.
84. MacKenna D, Summerour SR, Villarreal FJ. Role of mechanical factors in modulating cardiac fibroblast function and extracellular matrix synthesis. Cardiovasc Res 2000; 46:257-263.
85. Weber KT, Janicki JS, Shroff SG et al. Collagen remodelling of the pressure-overloaded, hypertrophied nonhuman primate myocardium. Circ Res 1988; 62:757-765.
86. Samuel J-L, Corda S, Chassagne C et al. The extracellular matrix and the cytoskeleton in heart hypertrophy and failure. Heart Failure Rev 2000; 5:239-250.
87. Simpson DG, Reaves TA, Shih D et al. Cardiac Integrins: The ties that bind. Cardiovasc Pathol 1998; 7:135-143.
88. Giancotti FG, Ruoslahti E. Integrin signaling. Science 1999; 285:1028-1032.
89. Ross RS, Borg TK. Integrins and the myocardium. Circ Res 2001; 88:1112-1119.

90. Ruwhof C, van der Laarse A. Mechanical stress-induced cardiac hypertrophy: Mechanisms and signal transduction pathways. Cardiovasc Res 2000; 47:23-37.
91. Ingber D. Integrins as mechanochemical transducers. Curr Opin Cell Biol 1991; 3:841-848.
92. Danowski BA, Imanaka-Yoshida K, Sanger JM et al. Costomeres are sites of force transmission to the substratum in adult rat cardiomyocytes. J Cell Biol 1992; 118:1411-1120.
93. Schlaepfer DD, Hanks SK, Hunter T et al. Integrin-mediated signal transduction linked to Ras pathway by GRB2 binding to focal adhesion kinase. Nature 1994; 372:786-791.
94. Cybulsky AV, Bonventre JV, Quigg RJ et al. Extracellular matrix regulates proliferation and phospholipid turnover in glomerular epithelial cells. Am J Physiol 1990; 259:F326-F337.
95. Ingber DE, Prusty D, Frangioni JV et al. Control of intracellular pH and growth by fibronectin in capillary endothelial cells. J Cell Biol 1990; 110:1803-1811.
96. Franchini KG, Torsoni AS, Soares PH et al. Early activation of the multicomponent signaling complex associated with focal adhesion kinase induced by pressure overload in the rat heart. Circ Res 2000; 87:558-565.
97. Maniotis AJ, Chen CS, Ingber DE. Demonstration of mechanical connections between integrins, cytoskeletal filaments, and nucleoplasm that stabilize nucleur structure. Proc Nat Acad Sci USA 1997; 94:849-854.
98. Wang YG, Samarel AM, Lipsius SL. Laminin binding to β_1-integrins selectively alters β_1- and β_2-adrenoceptor signalling in cat atrial myocytes. J Physiol 2000; 527:3-9.
99. Sharp WW, Simpson DG, Borg TK et al. Mechanical forces regulate focal adhesion and costomere assembly in cardiac myocytes. Am J Physiol 1997; 273:H546-H556.
100. Ahn AH, Kunkel LM. The structural and functional diversity of dystrophin. Nat Genet 1993; 3:283-291.
101. Ohlendieck K. Towards an understanding of the dystrophin-glycoprotein complex: Linkage between the extracellular matrix and the membrane cytoskeleton in muscle fibers. Eur J Cell Biol 1996; 69:1-10.
102. Kaprielian RR, Severs NJ. Dystrophin and the cardiomyocyte membrane cytoskeleton in the healthy and failing heart. Heart Failure Rev 2000; 5:221-238.
103. Kostin S, Heling A, Hein S et al. The protein composition of the normal and diseased cardiac myocyte. Heart Failure Rev 1998; 2:245-260.
104. Kaprielian RR, Stevenson S, Rothery SM et al. Distinct patterns of dystrophin organization in myocyte sarcolemma and transverse tubules of normal and diseased human myocardium. Circulation 2000; 101:2586-2594.
105. Hoffman EP, Kunkel LM. Dystrophin abnormalities in Duchenne/Becker muscular dystrophy. Neuron 1989; 2:1019-1029.
106. McNally E, Allikian M, Wheeler MT et al. Cytoskeletal defects in cardiomyopathy. J Mol Cell Cardiol 2003; 35:231-241.
107. Fowler VM. Regulation of actin filament length in erythrocytes and striated muscle. Curr Opin Cell Biol 1996; 8:86-96.
108. Sadoshima J, Takahashi T, Jahn L et al. Roles of mechano-sensitive ion channels, cytoskeleton, and contractile activity in stretch-induced immediate-early gene expression and hypertrophy of cardiac myocytes. Proc Natl Acad Sci USA 1992; 89:9905-9909.
109. Wang N, Butler JP, Ingber D. Mechanotransduction across the cell surface and through the cytoskeleton. Science 1993; 260:1124-1127.
110. Capetanaki Y. Desmin cytoskeleton in healthy and failing heart. Heart Failure Rev 2000; 5:203-220.
111. Yang X, Salas PJI, Pham TV et al. Cytoskeletal actin microfilaments and the transient outward potassium current in hypertrophied rat ventriculocytes. J Physiol 2002; 541:411-421.
112. Otey CA, Kalnoski MH, Bulinski JC. Immunolocalization of muscle and nonmuscle isoforms of actin in myogenic cells and adult skeletal muscle. Cell Motil Cytoskeleton 1988; 9:337-348.
113. Rybakova IN, Patel JR, Ervasti JM. The dystrophin complex forms a mechanically strong link between the sarcolemma and costameric actin. J Cell Biol 2000; 150:1209-1214.
114. Hilgemann DW. Cytoplasmic ATP-dependent regulation of ion transporters and channels: Mechanisms and messengers. Ann Rev Physiol 1997; 59:193-220.
115. Cantiello HF. Role of actin filament organization in cell volume and ion channel regulation. J Exp Zool 1997; 279:425-435.
116. Guharay F, Sachs F. Stretch-activated single ion-channel currents in tissue-cultured embryonic chick skeletal muscle. J Physiol 1984; 352:685-701.
117. Kim D. Novel cation-selective mechanosensitive ion channel in the atrial cell membrane. Circ Res 1993; 72:225-231.
118. Fuchs E, Cleveland DW. A structural scaffolding of intermediate filaments in health and disease. Science 1998; 279:514-519.

119. Milner DJ, Taffet GE, Wang X et al. The absence of desmin leads to cardiomyocyte hypertrophy and cardiac dilation with compromised systolic function. J Mol Cell Cardiol 1999; 31:2063-2076.

120. Balogh J, Merisckay M, Li Z et al. Hearts from mice lacking desmin have a myopathy with impaired active force generation and unaltered wall compliance. Cardiovasc Res 2002; 53:439-450.

121. Bloom S, Lockard VG, Bloom M. Intermediate filament-mediated stretch-induced changes in chromatin: A hypothesis for growth initiation in cardiac myocytes. J Mol Cell Cardiol 1996; 28:2123-2127.

122. Hill TL. Kirschner MW. Bioenergetics and kinetics of microtubule and actin filament assembly-disassembly. Int Rev Cytol 1982; 78:1-125.

123. Joshi HC, Chu D, Buxbaum RE et al. Tension and compression in the cytoskeleton of PC 12 neurites. J Cell Biol 1985; 101:697-705.

124. Cooper G. Cardiocyte cytoskeleton in hypertrophied myocardium. Heart Failure Rev 2000; 5:87-201.

125. Tagawa H, Wang N, Narishige T et al. Cytoskeletal mechanics in pressure-overload cardiac hypertrophy. Circ Res 1997; 80:281-289.

126. Yamamoto S, Tsutsui H, Takahashi M et al. Role of microtubules in the viscoelastic properties of isolated cardiac muscle. J Mol Cell Cardiol 1998; 30:1841-1853.

127. Tsutsui H, Ishihara K, Cooper G. Cytoskeletal role in the contractile dysfunction of hypertrophied myocardium. Science 1993; 260:682-687.

128. Tsutsui H, Tagawa H, Kent RL et al. Role of microtubules in contractile dysfunction of hypertrophied cardiocytes. Circulation 1994; 90:533-555.

129. Howarth FC, Calaghan SC, Boyett MR et al. Effect of the microtubule polymerising agent on contraction, Ca^{2+} transient and L-type Ca^{2+} current in rat ventricular myocytes. J Physiol 1999; 516:409-419.

130. Schaper J, Froede R, Hein S et al. Impairment of the myocardial ultrastructure and changes of the cytoskeleton in dilated cardiomyopathy. Circulation 1991; 83:504-514.

131. Heling A, Zimmermann R, Kostin S et al. Increased expression of cytoskeletal, linkage, and extracellular proteins in failing human myocardium. Circ Res 2000; 86:846-853.

132. Kostin S, Hein S, Arnon E et al. The cytoskeleton and related proteins in the human failing heart. Heart Failure Rev 2000; 5:271-280.

133. Zile MR, Green GR, Schuyler GT et al. Cardiomyocyte cytoskeleton in patients with left ventricular pressure overload hypertrophy. J Am Coll Cardiol 2001; 37:1080-1084.

134. Bers DM. Myofilaments: The end effector of E-C coupling. In: Bers DM, ed. Excitation-Contraction Coupling and Cardiac Contractile Force. 2nd ed. Dordrecht: Kluwer Academic Publishers, 2001:19-38.

CHAPTER 4

Mechanoelectric Transduction/Feedback:
Physiology and Pathophysiology

Max J. Lab*

Abstract

Cardiac "mechanotransduction" involves various physiological and biophysical phenomena in which mechanical energy is transduced to changes in function of cardiac myocytes and of the whole heart. In this chapter different manifestations of mechanotransduction are reviewed, with special emphasis on the "mechano-electric" feedback aspect. The chapter covers both physiological and pathological roles of mechanical stimulation of heart tissue.

Brief Historical Introduction

Several long-standing studies have shown that a mechanical change in heart can produce an electrical change. The easiest one to appreciate, and one well known in physiology, is the "Bainbridge reflex".[1] A simple hydrodynamic (mechanical) event, a raised venous return, raises heart rate. This phenomenon is more than likely via stretch of the sinoatrial node enhancing spontaneous depolarisations (e.g., see Fig. 3C). Mechanically induced electrical changes were also described in Bozler's[2] early study and in Penefsky and Hoffman's experiments.[3] Deck[4] followed the depolarisation tack. Still following this theme, stretch can affect Purkinje fibres, as demonstrated by Dudel and Trautwein,[5] and Kaufman and Theophile.[6] The latter authors also showed, crucially, spontaneous depolarisations in stretched papillary muscle (Fig. 1A). Von M. Stauch[7] a practising physician, was one of the first to show a mechanically induced change in repolarisation in frog heart on changing the mode of contraction, from auxotonic to isovolumic. Also in an almost identical preparation, Lab in a pilot study[8] introduced the possibility that mechanoelectric transduction can operate in tissue other than that more obviously dedicated to transducing mechanical signals to electrical signals for physiological purposes (e.g., auditory hair cells, pacinian corpuscles, proprioceptors and muscle spindles - he asked the question "Is there mechanoelectric transduction in heart?").

"Mechanotransduction" describes a broad based biophysical phenomenon in which mechanical energy is transduced to a different form of energy. "Mechanoelectric transduction" applies when a mechanical stimulus is transduced into an electrical signal. In cardiac muscle, the term "Feedback" is sometimes used, for it very likely underlies a regulatory-biophysical control, and a pathophysiological role.[9-24] That is, the phenomenon appears to have a fundamental role in cardiac physiology and pathophysiology. This is not unexpected, for the heart's elemental physiological function is mechanical. Moreover, as most of its pathology ultimately invokes some mechanical derangement or other, it is not surprising also, that mechanoelectric transduction/feedback rears up in protean guises.

*Max J. Lab—Imperial College, National Heart and Lung Institute, London, U.K.
Email: m.lab@imperial.ac.uk

Cardiac Mechanotransduction, edited by Matti Weckström and Pasi Tavi.
©2007 Landes Bioscience and Springer Science+Business Media.

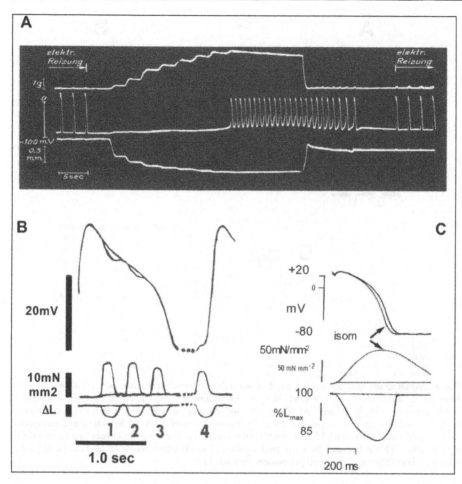

Figure 1. Electromechanical records from isolated superfused ventricular preparations. A) Sustained stretch, indicated in top and bottom traces, produces progressive diastolic depolarisation in the microelectrode recording, middle trace, until spontaneous action potentials are generated (reprinted with kind permission of Springer Science and Business Media).[6] B) The first two transient stretches, indicated in the bottom two traces, produce transient repolarisation potentials in the plateau phase of the superimposed action potentials (insulated gap - top traces). Stretch 4 near diastole produces transient depolarisations reaching threshold, whereas stretch 3 produces no deviation. That is, stretch 3 could be at an effective stretch reversal potential (reprinted with permission from ref. 60). C) Compared with lightly loaded contraction, heavily loaded isometric (isom) contraction reduces the action potential duration (top traces) (reprinted with permission from ref. 44). A and C) mammalian papillary muscle, B) amphibian ventricular strip.

Prevalence[16,19]

Mechanoelectric transduction/feedback expresses in many living cells,[25] and may be found in several cardiac ventricular experimental preparations. Mechanically induced changes in electrophysiology have a type of phylogenetic prevalence being found in many species using a variety of simultaneous electromechanical recordings (for e.g., see Figs. 1-5). The electrophysiological changes are found in the membrane patch (diagrammatically indicated in Fig. 2A), whole cell current (Fig. 2B), single cell (Fig. 2C), ventricular strip (Fig. 1B), papillary muscle (Figs. 1A, C; 5B), mammalian heart (isolated perfused, and intact heart in situ as well as in

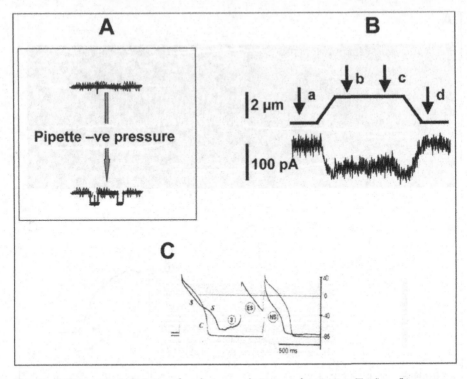

Figure 2. A) Diagrammatic depiction of patch currents during stretch activation. Top "trace", no currents. After stretch (suction via patch pipette) the open probability of stretch activated channels "increase" - downward deflections in "record". B) Whole cell patch current observed with stretch (a to b), which is reversible on de-stretch (c to d) (reprinted with permission from ref. 226). C) Stretch (s) of a ventricular myocyte, compared with control (c) accelerates early plateau repolarization but crosses over to later delay repolarization. This is followed by a late afterdepolarization (3) which seems to reach threshold for a premature beat (ES) (reprinted with permission from ref. 114).

man (Fig. 3A and B). Figure 3 also shows records of signals from intact atrium-sinoatrial node (Fig. 3C) and atrial tissue (Fig. 3D). Virtually all the anatomical subdivisions of the heart (Fig. 4A), cells as well as intact chambers, show aspects of mechanoelectric transduction or feedback.[9-21,26] The preparations include atrial fibroblasts,[27] sino-atrial node[28] (see also, an interesting review ref. 29) and atrial tissue.[12,22,30-32]

Similarly, virtually all the preparations studied show concordant electrophysiological expression of mechanically induced electrical changes - a type of "Electrophysiological" Prevalence (Fig. 4B). Briefly reiterated, membrane distortion increases channel opening probability (Fig. 4B @ 8 o'clock going to 9 o'clock) the likely basis of the observed mechanically induced diastolic and threshold depolarisations. Generally the action potential associated with a large load is short (Fig. 4B starting @ 6 o'clock going through centre to AP @ 11 o'clock, and MAP @ 1 o'clock) compared with a contraction having a reduced load (solid lines). This can be associated with an "early afterdepolarisation" (1 o'clock).

The electrocardiogram (ECG) is the recording of surface potentials and is the manifestation through a volume conductor of electrical vectors derived from the inhomogeneous spread of cellular action potentials through the heart, and the ECG expresses mechanoelectric feedback.[16] The QT interval of the ECG prolongs with reduced load (Fig. 4B ECG @ 2 o'clock, going to 4 o'clock). The T-wave of the electrocardiogram is, more directly than is the QT interval, some

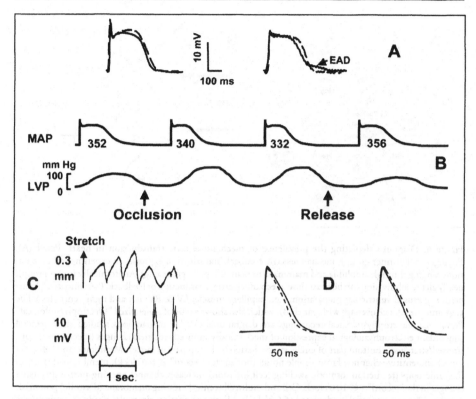

Figure 3. A) Superimposed monophasic action potentials from in-situ right ventricle of sheep. Outflow obstruction shortens the action potential duration early on (solid lines). Some regions show a cross-over and an early afterdepolarization (EAD) (reprinted with permission from ref. 227). B) Monophasic action potentials (MAP) from left ventricle in man. Outflow tract occlusion increases the left ventricular pressure (LVP) and reduces the action potential duration. The changes are reversible on release (reprinted with permission from ref. 69) C) Monophasic action potentials (lower trace) and segment motion (upper trace) from the sinus node area in intact pig heart in situ. The preparation was pharmacologically denervated. Superimposed on the contractile activity are the slower mechanical excursions with respiration. The increasing stretch over the first 3 action potentials show increasing slopes of the 3 preceding diastolic depolarisations, and a relatively rapid heart rate. Decreased stretch is associated with a flatter slope in diastolic period 4 and the inter-beat interval is long (reprinted with permission from ref. 127). D) Superimposed monophasic action potentials from left atrium of isolated Langendorff guinea-pig heart. Stretch produces 2 responses (dotted action potentials) - shortening (left panel, and lengthening (right panel) (reprinted with permission from ref. 26).

function of electrophysiological inhomogeneity of repolarisation (roughly epi to endocardium and base to apex - opposite in route to depolarisation). This produces similar general directions of the resulting vectors, creating the anomalous "upright" T-wave of the ECG. As intramyocardial contraction is also inhomogeneous it is possible that mechanoelectric feedback can play a modulatory role. Mechanical loading does change the T wave[33] (Fig. 4B ECG 2 o'clock going to 4 o'clock). This may be the result of the volume change modulating regional mechanoelectrical inhomogeneity.[34] There are, however, alternative purely physical, explanations for T wave changes with volume changes.[35] The U wave of the ECG may be modulated by mechanical changes[36] and also, they may be related to early afterdepolarisations.[37]

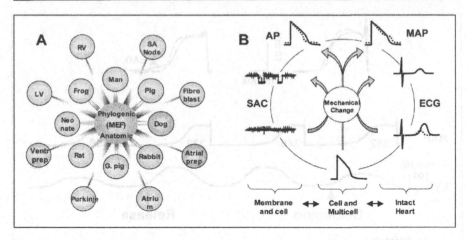

Figure 4. Diagrams depicting the prevalence of mechano-electric transduction in heart. Panel (A), Phylogenically (inner group), mechanoelectric feedback/transduction expresses in experimental preparations ranging through amphibia and mammals, to man. (G. pig = guineapig). Anatomically (outer group), nearly all the subdivisions of the heart show mechanoelectric transduction or feedback. (SA: sinoatrial; Ventr prep: superfused ventricular preparation, e.g., papillary muscle; LV & RV: left and right ventricles.) The anatomic prevalence overlaps with panel (B), which also shows a type of generalized "Electrophysiological" Prevalence (electrophysiological recordings are diagrammatic). The preparations studied show general concordant electrophysiological expression of mechanically induced electrical changes (mechanically unstressed/stretched - bottom part of circle; stressed/stretched - top part of circle). Membrane and cell (left) - mechanosensitive channel. From simple noise, (horizontal "record" @ 8 o'clock), membrane distortion (i.e., micropipette suction, osmotic swelling, cell inflation) increases channel opening probability (step changes in the "record" - @ 10 o'clock. Cell and multicell (center) - resting membrane (diastolic) potential. A normal "action potential" is depicted @ 6 o'clock. Micro and suction electrodes produce, respectively, action potentials (AP) and monophasic action potentials (MAP) depicted @ 11 and 1 o'clock. Continuous or transient stretch depolarises the myocardium (dotted "records"). Action potential. A large load shortens the cellular AP (dotted "record" @ 11 o'clock). In intact heart - right, the ventricular MAP (@ 1 o'clock) shows comparable changes (dotted MAP) including afterdepolarisations. The electrocardiogram (ECG). The normal ECG (@ 4 o'clock) after changing load, shows QT interval and T wave changes (dotted "ECG" @ 2 o'clock - also, see text). Reprinted with permission from Elsevier.[228]

In general, the consistency of the observations seems remarkable but there are some apparent anomalies. Although a high load usually shortens repolarisation, some studies have demonstrated that it could lengthen. The likely explanation is that an early afterdepolarisation, which occurs at the tail end of the action potential, prolongs the action potential. An associated U wave slurring the QT interval could produce apparent QT prolongation. The inconsistencies could be apparent rather than real, although one study in intact atrium shows both prolongation and reduction in monophasic action potential, but in different regions.[26]

The phylogenic, anatomic, and electrohysiological prevalence imply physiological roles.[14,16] First, heart rate. The sinoatrial node controls normal heart rate, and, as long established in physiology texts, change in pressure in even in denervated right atrium is one overall modulator of heart rate. This includes a local direct mechanical modulation of sinus arrhythmia (see Fig. 3C), which is usually exclusively explained by external and central reflexes. Second, there is mechanoelectric modulation of mechanical function. Homeometric autoregulation in the intact heart describes the situation where a slow rise in developed pressure accompanies an abrupt increase in afterload. This ("Anrep" effect) has been explained by variations in regional ventricular blood flow,[38] with a slow recovery of an initial subendocardial ischaemia. However, mechanically induced changes in action potential and/or intracellular calcium may offer an alternative explanation.[39]

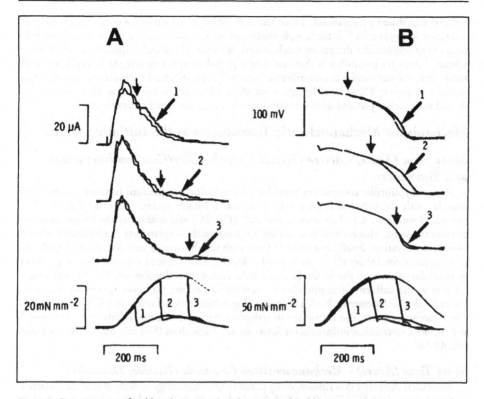

Figure 5. Consequences of sudden shortening (quick tension release) during isometric contraction of cat papillary muscle. The mechanical interventions are superimposed in the lower traces. Release at 3 different times 1, 2, and 3 are indicated by the vertical arrows, with the diagonal arrows indicating the effects on calcium panel A and action potential panel B compared with control. The intracellular calcium transients and action potential (top 3 traces in each panel) rise with each mechanical tension release. The action potential rise is in effect a prolongation of action potential duration (reprinted with permission from ref. 44).

Some Pitfalls in Interpretation

Mechanically-Induced Electrical Artefact

The conception of mechanoelectric feedback, particularly in the intact heart, has been traumatic. The methodology includes simultaneous impositions or measurements of mechanical variables (force, length, pressure), and electrophysiological variables (probability of channel opening, intracellular action potentials, monophasic action potentials, electrocardiogram). Any mechanical change in the experimental preparation will, physically, alter the relationship between the electrical generator and the recording electrode. This then raises the spectre of mechanically induced electrical artefact. This would apply to all the preparations; from membrane patch, with a change in the seal, to the electrocardiogram with an altered electrical vector. However, the results show a remarkable concordance in the different experimental situations in that mechanically related changes can induce electrically related changes that are generally analogous. Moreover, an influential observation suggesting that the observations are real is that the mechanically-induced electrophysiological changes are often associated with those underlying the generation of premature beats and arrhythmia, and the latter is often seen (as reviewed refs. 9,10,13,40). However, the literature reports the existence of stretch induced arrhythmia in intact ventricle more often than it reports mechanically induced arrhythmia in isolated superfused preparations. There seems to be only one study showing stretch induced activation in single cells.[41] If the sample studies are representative, other factors may operate in intact ventricle besides the above mechanisms, for example stretch-induced catecholamine release.[42] Another possibility is that ventricular global stretch activates the Purkinje network rather than the ventricular contractile cell, because Purkinje diastolic depolarisations are more sensitive to stretch. However although, most of the observations mitigate against artefact, one should nonetheless proceed with caution in interpreting any specific experiment.

Mechanisms: Mechanoelectric Transducers as the Initiating Event

Short Term (Acute) - Active (Systolic) Myofibrillar/Calcium Interaction as a Transducer

Active myofibrillar contraction provides a transducing mechanism. Isotonic (shortening) muscle, with the reduced force production, has a longer action potential duration[43,44] compared with a stretched or isometric muscle (Fig. 1C) and it also has the longer duration calcium transient. This is counter-intuitive: but a reasonable explanatory mechanism is related to force deactivation. Briefly, the reduced force with the shortening muscle reduces the affinity of troponin c for calcium[44,45] (see also refs. 46,47). This reduced calcium buffering allows intracellular calcium to rise in the face of a reduced force production (see Fig. 5). The raised calcium would influence transmembrane currents to prolong the action potential duration. For example via electrogenic Na/Ca exchange (Fig. 6 bidirectional arrow in 3 o'clock direction. These calcium aspects, also important in arrhythmogenesis (see below), have been reviewed[39] and includes mechanical influences on Sarcoplasmic Reticulum (SR) calcium cycling (see also refs. 48-50).

Short Term (Acute) - Mechanosensitive Channels (Passive, Diastolic)

Since their definitive description, using patch electrophysiology in skeletal muscle,[51] stretch activated channels (SACs, Figs. 2A, 4B @ 9 o' clock, and 6 @ 2 o'clock) have been described in many tissues (for reviews see refs. 52-56) including heart.[57,58] Membrane stretch increases the open probability of the channels to admit charge-carrying ions, which influences the membrane potential.

Micropipette patch studies start with a residual membrane tension that may not be commensurate with that found in multicellular preparations, and multicellular preparations experience more tension than isolated cells (see also ref. 59). This questions the physiological validity of patch studies as applied to the intact heart. However, characteristics of the channel electrophysiology have found equivalence in intact heart, where monophasic action potentials (MAPs) can be used to gain qualitative insights into cellular electrophysiology: for example isolated perfused heart of frog,[60] and mammals.[61-63] Notably, there is a type of "stretch/volume" reversal potential,[60,64] where a stretch early in the action potential produces a repolarising tendency, whereas if late it produces depolarisation. There is no change at the "reversal potential". There are other examples of the equivalence, in different guises, in intact heart in situ in experimental preparations (see previous; and also refs. 9,33,65-68) and in man.[69] Load induced reductions in action potential duration, in these examples, can be explained by SACs. Another example is that SAC blockers, such as gadolinium[61] and streptomycin[70-72] in intact preparations are commonly used to block stretch induced electrophysiological changes. However, the results may have to be carefully interpreted. Streptomycin's effects as a blocker appear heterogeneous in intact atrium.[26]

The ATP activated potassium (KATP) channel subunits (Kir6.1, Kir6.2, SUR1A, SUR1B, SUR2A, and SUR2B)[73] have been found in atria from neonatal rats. The KATP channel is attached to the cytoskeleton, and it has been suggested to be mechanosensitive.[74] Several other channels appear to be mechanosensitive, including the L type Ca channel[75] sodium channels[76] and the tandem pore potassium channels such as TREK-1 (indicated in Fig. 6 @ 8 o'clock),[77]

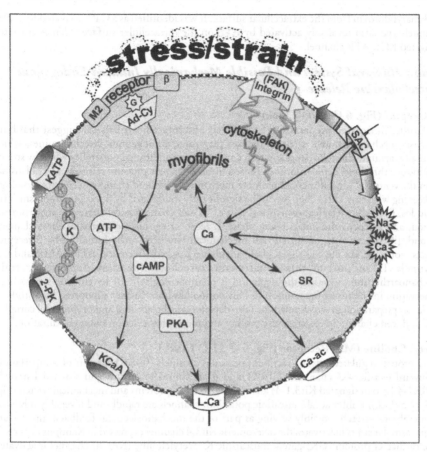

Figure 6. Diagram of some of the cellular transducers and mechanisms initiating mechanoelectric feedback/transduction. Mechanical stress/strain (top) is transmitted to the cell "membrane" via the extracellular matrix depicted by the stippled cap of banners. The mechanical changes also reach the cell interior and intracellular cytoskeleton (@ 1 o'clock) via the integrins and focal adhesion kinases (FAK) to modulate many of the transducer mechanisms. The various mechanosensitive channels and receptors, which can influence electrophysiology, are diagrammatically distributed in the membrane. For example, stretch activated channels (SAC) - @ 2 o'clock) open to admit Na^+ and Ca^+. The former influences the electrogenic Na-Ca exchange (@ 3 o'clock) to influence the action potential, and the latter raises intracellular Ca^+ also via loading the sarcoplasmic reticulum (SR). Other mechanosensitive channels, which can acutely influence electrophysiology, include the KATP channel (@ 10 o'clock); the L type Ca channel (LCa - @ 6 o'clock); the tandem pore potassium (2-PK - @ 8 o'clock). Mechanical change can influence some receptors and their G proteins (G). β receptor activity (@ 11 o'clock) via cAMP, adenyl cyclase (Ad-Cy), and ATP. (The mechanosensitive muscarinic G protein gated (M2) ACh receptor is nominally depicted at 9 o'clock, near intracellular potassium - K). There are also mechanically-induced changes in ATP which can influence other channels, including a stretch activated KCa,ATP channel (KCaA) - @ 7 o'clock), and the tandem pore channel, TREK-1 (incorporated "2-PK" - @ 8 o'clock). Myofibrillar mechanical changes (center) affect intracellular Ca^+ is pivotal. In addition to influencing Na-Ca, it can impinge on some of the previous channel types, as well as other calcium-activated channels (Ca act - @ 5 o'clock). Reprinted with permission from Elsevier.[228]

which appears not to involve the cytoskeleton - and IKAA,[78] ion exchangers, (Fig. 6 @ 3 o'clock).[79] In cultured chick ventricular myocytes, a stretch-activated ion channel[80] was identified as a high-conductance K^+-selective channel blocked by gadolinium, tetraethylammonium, and charybdotoxin from the extracellular surface. It was identified as a Ca^{2+}-activated K^+ (KCa) channel type, also reversibly activated by ATP on the intracellular surface. That is, a stretch activated KCa, ATP channel.

Neuro-Humoral System and Possible Mechanically Induced Endogenous Catecholamine Release

β Receptor (Fig. 6 @ 11 O'Clock)

Studies in isolated preparations (no nerves) and intact heart in-situ, suggest that both intrinsic mechanisms, and extrinsic reflexes play roles. Endogenous catecholamines, which naturally facilitate diastolic depolarisation constitute one additional possibility whereby stretch can induce threshold depolarisations. Raised intraventricular pressure releases catecholamines from the ventricle[42] and this could promote mechanically induced changes in electrophysiology. In keeping with this, these changes can be significantly modulated by β receptor agonism. This includes, first, the electrical restitution curve.[67] Load changes accentuate the supernormal period, and steepens the initial rising phase. β receptor agonism exacerbates these changes. Second, β receptor antagonism curtails mechanically induced arrhythmia.[81] The β receptor effects would be via the cell signal chain involving β agonist/receptor, ATP, cAMP, and Ca. channels. In support of this signal chain, catecholamine store depletion curtails mechanoarrhythmia whether the depletion is pharmacological[82] or by chronic sympathetic denervation in the intact preparation.[83] This notion has been further supported by a study in the intact preparation in which reserpine (also depletes catecholamines) and propranolol curtailed load-induced changes in electrical excitability and monophasic action potential duration.[84]

Acetyl Choline (M2) Receptor (Fig. 6 @ 11 O'Clock)

G-protein regulated inward-rectifier potassium channels (GIRK) are part of a superfamily of inward-rectifier K^+ channels. GIRK1 and 4 are found in atrium and sinoatrial node.[85] GIRK1-4 (also designated Kir3.1-4) are modulated by G-proteins and mechanical stretch (Fig. 6 @ 11 o'clock). Rabbit atrial muscarinic potassium channels are rapidly and reversibly inhibited by membrane stretch, possibly serving as part of the mechanoelectrical feedback interaction. Hypo-osmolar stress inactivates the heteromeric Kir3.4 channel expressed in Xenopus oocytes.[86] The cardiac G protein (GK)-gated, muscarinic K^+ acetylcholine (KACh) channel is activated by stretch if ACh is present and this appears to be independent of receptor/G protein, probably via a direct effect on the channel protein/lipid bilayer.[87]

A Note on a Possible Mechanically Mediated Crosstalk (MMC)

The foregoing indicates that mechanosensitivity is protean in its expression, interacting among many channels and signals, and at different levels - demonstrating crosstalk. This has been briefly alluded to.[16] Some licence is taken in characterizing this "crosstalk" as being a mechanical component, which provides a common link between the various interactions at molecular, cellular, and macro levels.[88,89] With this licence one may call it a type of mechanically mediated crosstalk (MMC).

The Cytoskeleton's Role "Tensegrity": A MMC Vehicle?
(Fig. 6 1-2 O'Clock Direction)

Mechanotransduction appears to involve, or to be modulated by, the intracellular cytoskeleton,[89] and its insertions into the surrounding cell membrane via focal adherence complexes. From the electrophysiological aspect the cytoskeleton may be important because, physiologically and geometrically, parts of the cytoskeleton are associated with ion channel function.[55,90-92] Many stimuli (some indicated in Fig. 6) share intracellular signal pathways with mechano-electrochemical ones, and this can produce composite responses. Here the role of the intracellular cytoskeleton appears to be crucial. It can form the mechanical engineering and biochemical basis for intracellular and intercellular crosstalk. Any membrane channel associated with the cytoskeleton could provide a candidate for crosstalk. Stress-strain could be transmitted to the channel via the cytoskeleton to activate it, or the cytoskeleton could somehow shield the channel from the mechanical transmission of strain via other paths. The ATP activated potassium (KATP) channel, which is mechanosensitive, is attached to the cytoskeleton. Several ion channels are regionally fixed in the membrane,[93] and their modulation by the cytoskeleton is also possible.

Under physiological conditions, the mechanical interconnecting system is prestressed, conferring mechanical balance and stability. This has been described as "tensegrity",[89] and there is thus a clear line of force communication from the extracellular matrix to intracellular cytoplasmic structures.[88] Mechanical stimuli can provoke changes in adhesions between the cell and the extracellular matrix via the fibronectin and collagen networks. In this way, mechanically distorting one cell can raise intracellular Ca^{2+} of its neighbouring cell.[94]

Longer Term: Mechanotransducers via Cell Signals Cascades

The elaborate adhesion complex of the cytoskeleton[95,96] also contains focal adhesion kinase (FAK - Fig. 6 @ 1 o'clock). This kinase can be the initiator or modulator of many downstream cell signals and kinases, including tyrosine kinase[97] and the renin angiotensin system as reviewed in ref. 98. These are possible/probable mechanisms whereby chronic stretch can produce remodelling and hypertrophy.

Mechanical stimulation also releases a multitude of other second messengers[39,98] that include cAMP (Fig. 6 off centre, left). More detailed individual mechano-chemo-sensitivity, including ionic currents carried by channels other than specific stretch-activated ones, has been reviewed.[55] Transducers could be provided via other receptors—see also brief overviews specifically related to mechanoelectric transduction[16]—which could affect intracellular Ca, first, via IP3 and SR Ca release. Second, through PLC to facilitate membrane translocation to influence Na/H exchange, and so intracellular Ca via Na/Ca exchange (Fig. 6 at 3 o'clock). These calcium changes thus produce electrical changes.

Theoretical (Abstract) Reflections

An Integrative Mechanism

Cardiac mechanotransduction occurs at different levels, and there is a possibility that mechanoelectro-chemical transduction forms part of a network of mechanically mediated crosstalk (MMC) in heart. The crosstalk tracks a path from mechanosensor to the whole heart within an integrative system. It spans other regulatory systems and processes. Finally, if elements of the normal crosstalk experience derangement it contributes to cardiovascular disease, and potentially lethal arrhythmia.

Previous reviews have suggested that mechanoelectric transduction/feedback may form part of an intracellular myocardial stabilising system between mechanical and electrical events, and mechanoelectric transduction or feedback has been described as an integrating regulatory control system, analogous with other regulatory systems (see refs. 14,16). As with the other mechanisms (e.g., neurendocrine system), it fulfils or confirms some criteria in an analogous fashion. For example, it operates at several levels, from the molecular to the integrating system: it displays homeostatic self regulation, and finally, provides a conduit for pathology. In fulfilling these, first, it tracks a path from mechanosensor to the whole heart within the integrative system. The outline under the previous heading, "*Prevalence*", incorporates this path. Second, there is considerable "crosstalk" between the various mechanosensitive mechanisms, often with intracellular calcium changes providing some additional linkage to the mechanical linkage. Third, it interacts with the other regulatory integrative systems and processes, or that at least, potential mechanisms exist for this integration. For example, in the intact heart, autonomic factors appear to interact with mechanoelectric feedback (see above). Fourth, as both neural and endocrine control systems are well described by negative (and sometimes positive) feedback loops, mechanoelectric feedback, as its epithet implies, it is amenable to scrutiny in this form. Finally, as with the other systems, when it malfunctions, it contributes to cardiovascular disease.

Mechanically Mediated Crosstalk (MMC) in the Intact Heart: Regulation of Cardiac Function

Spatially, the integrative stress/strain system transmits to the whole heart. This would be by cell attachment through the adherence proteins to the extracellular matrix, which connects cells to each other. Forces at one end of a group of cells are thus transmitted throughout the organ, invoking tensegrity.[89] This provides a clear mechanism by which a mechanically induced functional crosstalk can occur in the intact heart. Additionally, the heart is part of a hydrodynamic system. MMC in the intact heart could be via a mechanical, hydraulically mediated mechanism. That is, pressure volume changes in a system containing incompressible blood could provide the mechanically mediated crosstalk. Systemic arterial pressure, and or venous volume changes can influence cardiac electrophysiology and function.[14,33,40,55,65,66,99-101] This would invoke the variety of cellular mechanisms covered above.

We know that stretch can raise heart rate by mechanisms other than neural reflexes. Sinoatrial stretch in isolated perfused hearts increases heart rate (ref. 1), probably by a mechano-electric mechanism[102] working through stretch activated channels and increasing the slope of the diastolic depolarisation. Thus, in addition to nervous reflex mechanisms, cellular mechanoelectric transduction provides integrative control mechanisms for raising heart rate in response to an increased venous return to the heart.

On a marginally different track, one of the major mechanosensitive regulatory mechanisms in heart is related to contractile function per se. Myocardial stretch increases force production. This has been well studied, including the pioneering work of Frank, Starling, Patterson in the intact ventricle. The prevailing dogma for the familiar Starling's Law of the Heart is a length/force sensitivity of the contractile proteins - perhaps related to calcium.[103,104] A membrane mechanosensitivity probably also makes a contribution in the intact ventricle by mechanotransduction, possibly via stretch activated channels.[82] Studies at the membrane level corroborate this, showing that stretch can raise intracellular calcium by affecting mechanosensitive channels,[94,105,106] and so increase force.

Feedback[16]

Why use the term *"feedback"* (*"mechanoelectric feedback"*) at all, rather than *"mechanoelectric coupling"* or *"mechanoelectric transduction"*? One always regarded excitation contraction coupling as a unidirectional phenomenon (Fig. 7A, 1). Similarly, we could construe mechanoelectric coupling or transduction as being unidirectional (Fig. 7A, 3). The heart cyclically and regularly changes its mechanoelectric state. This is largely parallel in time and anatomical "space". Moreover, common processes are used in excitation coupling and mechanoelectric transduction. These processes provide a matrix for feedback. Interaction between ECC and mechanoelectric feedback (Fig. 7A) would produce fine-tuning of the regulatory processes in myocardium. In this way, any mechanical change would influence membrane electrophysiology and excitation (Fig. 7A, 3), and hence, in turn, mechanical function (Fig. 7A, 4). Although the term "Mechanoelectric Feedback" lacks rigor, it could partly describe its functional role. It also lends itself to the concept that the feedback loops could provide stability during electrical or mechanical perturbation under normal circumstances, but may destabilise the situation under pathological conditions. That is, disturbances in this tuning by pathological processes that induce mechanical changes in the heart should produce clinical syndromes that have in the past been difficult to explain, or treat, on a purely electrophysiological basis.

To start with, it can be part of a regulatory control system in intact heart. During contraction of the normal intact ventricle, myocardial segments show reasonably synchronous electrical and mechanical activity. Myocardial segments, in parallel, and in series have physiological time relationships. Dispersion of repolarisation is the difference in duration between APD in one

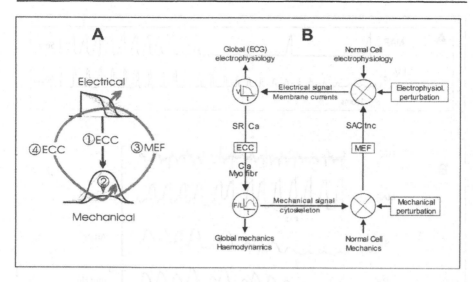

Figure 7. Mechanoelectric feedback (MEF) portrayed as simple feedback control loops. A) Simple biological portrayal. 1) In the "electrical" signal, the diagrammed short action potential via excitation contraction coupling (ECC) produces the taller contraction. For some reason (2) contraction reduces. This feeds back (3) via MEF to lengthen the action potential. This prolongation would via the next ECC cycle (4) increase the small mechanical signal. B) Simple engineering portrayal. In the top left corner the action potential, which is voltage vs time (V vs t) electrophysiologically expresses globally (up arrow) as the electrocardiogram (ECG). It also releases Ca$^+$ (direction down) from the sarcoplasmic reticulum (SR) for excitation contraction coupling (ECC). Ca interacts with myofibrillar proteins (Myofibr), produces contraction, in the bottom left corner, which is force or length against time (F/L vs t). This produces (down arrow) the global mechanical and haemodynamic changes, e.g., pressure/volume changes. The mechanical signals (via the cytoskeleton) including normal mechanical contraction (bottom horizontal arrow) feed in to a comparator (bottom right crossed circle). Any mechanical perturbation would also go to this comparator for summing. The signal out (direction up) then invokes MEF via mechanoelectric transducers, e.g., stretch activated channels (SACs) or the troponin c (tn-C) mechanism. The mechanically activated currents feed another comparator (top right crossed circle) summing with normal electrophysiological signals and any electrophysiological perturbation. This produces a membrane voltage signal to affect the action potential (returning to the top left corner). Reprinted with permission from Elsevier.[228]

area or segment and another. It is small but normally enough to give the vectors for the T wave of the ECG. This could conceivably have some relationship to mechanical dispersion via mechanoelectric feedback.[33] Moreover, the interaction between electrical and mechanical events operates within boundaries, and this relationship may have to be maintained for reasonable physiological and integrative function. Any disturbance in the system could destabilise matters, if not mitigated, to produce pathology (see below) - this instability can be a hallmark of a control system with severely altered loop gain. A highly simplified application of control theory (Fig. 7B) demonstrates that the interaction between the electrical and mechanical events can be depicted as a feedback control system. This keeps cellular electromechanical relationships (globally manifest in the ECG and pressure/volume) within normal boundaries.

The application of control theory here is vestigial, and some of the components in Figure 7B need distinguishing. It is difficult to precisely identify these in such a system. They include; the detailed controlled variables, the "comparator" (error detector) receiving the ME-feedback loop, and the "error signal" (difference between the feedback signal and the control signal (situation), which goes to a "controller". Additionally, the transfer function between current/voltage (I /V) and mechanical performance, as well as the transfer function between excitation and contraction need defining. Breakdown of the control system, or positive feedback results

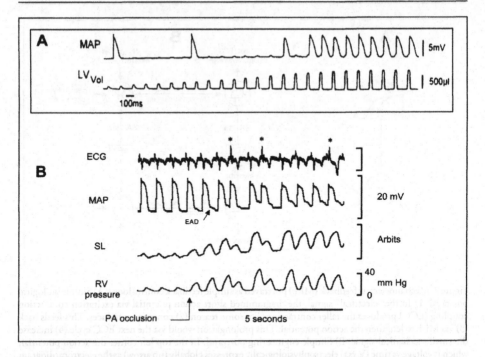

Figure 8. Mechanically induced ventricular contractions. A) Isolated heart in which progressively increasing inflation volumes (lower traces) produce progressively increasing depolarisations in the monophasic action potential (upper traces) from ventricle, until they reach a level in which each elicits a beat (reprinted with permission from ref. 64). B) Intact right ventricle (RV) in situ Pulmonary artery (PA) occlusion progressively increases RV pressure, and segment length (SL). The latter before occlusion show upward deflections (filling and stretch) with the downward deflections systolic ejection shortening. The monophasic action potential (MAP) has a normal configuration. PA occlusion produces paradoxical SL motion in that the SL bulges (stretches) in systole. This induces early afterdepolarisations (EAD) which appear able to reach threshold for premature beats (asterisks in the ECG trace) (reprinted with permission from ref. 227).

in altered physiology, or pathophysiology. It could be that breakdown of the control system produces the facilitation of mechanoelectric transduction which is becoming increasingly evident in myocardial pathology.[66,107-114]

Mathematical Advances

Several mathematical modelling approaches, using equations derived rigorously from experimental data, to mechanoelectric feedback have appeared, consolidating the electrophysiological studies.[17,115-119] Some modelling is beginning to consolidate pathophysiological studies of mechanoelectric feedback inducing arrhythmia,[120] and how mechanical heterogeneity can produce the matrix conducive to arrhythmia.[121] (see below).

Mechanoelectric feedback as a feedback process acting as regulatory system (Fig. 7), lends itself to nonlinear dynamical modelling.[16] This may have bearing on the stability of the system, the generation of "chaos" and/or cardiac arrhythmias. This nonlinear dynamical approach ("chaos") is being applied to biological rhythms, and has been argued and applied to mechanoelectric feedback embracing normal and pathophysiological situations.

- Mechanoelectric feedback has been theoretically argued. The feedback could stabilise the system, as chaotic processes may confer physiological adaptability.[122]

- Other models suggest a mechanical role in cardiac defibrillation.[123]
- Normal heart rate variability (HRV), and cardiac nonlinear dynamics can be influenced by respiratory sinus arrhythmia.[124] Ventricular dilatation allies with myocardial failure which can show chaotic dynamics[125] and a reduction of HRV, and there is loss of variability before demise.[126] One laboratory[127] has found that in beta blocked intact preparations mechanically shielding the sinus node from cyclic mechanical changes during respiration dramatically reduces respiratory sinus arrhythmia. Conceivably, the high end-diastolic pressures, and comparable mechanical changes in the right atrium during heart failure could persistently stretch the sinoatrial node making it less liable to cyclic respiratory stretch, and so reduce HRV.
- Parasystole and bigeminy may employ chaotic processes, and these arrhythmia can be mechanically induced.
- Electrophysiological alternans can be regarded as a period 2 doubling bifurcation, which may be a route to chaos. This alternans can precede ventricular fibrillation,[128] its heterogeneous nature contributing to this. Mechanical alternans, also prognostically bad, can be heterogeneous under pathological situations.[129] The inhomogeneous mechanical alternans, via mechanoelectric transduction, would produce electrical inhomogeneity, and arrhythmogenic electrical dispersion.
- Electrical restitution curves that show supernormality, with steepening of the rising phase, could provide a model of deterministic chaos, and these findings are in keeping with mechanically induced changes in restitution curves.[68] (Conversely, flattening restitution curves is antiarrhythmic.[130])

Pathophysiology: Clinically Related Expression

Mechanical perturbations can give rise to clinically relevant premature beats in ventricular myocardium - isolated and intact in situ; left and right ventricles (Figs. 1A and BC, 2C, 8B), as well as in atria (reviewed in ref. 12) (Fig. 9). MEF is increasingly being highlighted as a possible cause of sudden cardiac death in man, which is often arrhythmic (reviewed see refs. 10,21,26). Although this section discusses arrhythmia mainly in relation to the ventricle, atrial fibrillation is the commonest arrhythmia in man (and horses!) and is a potent source of morbidity and mortality (see a concerted effort at reviewing the problem).[131] However, the mechanisms and correlates that apply in the ventricle, can translate to the atrium. Atrial dilation appears to be a contributory factor to atrial fibrillation[12] as has also been demonstrated in experimental situations (Fig. 9). Importantly, pathological situations appear to amplify mechanoelectric feedback, see below.

In the clinical context, the phenomena described above have been demonstrated during acute stretch on intact heart but chronic stretch underlies many cardiac diseases. In these conditions the cytoskeleton, which plays a major role in modulating force transduction in cardiac cells is profoundly altered.[132,133] A close relationship between cytoskeleton and ionic channels[91,134-136] has been demonstrated which could contribute to the early action potential lengthening in heart failure and dilated cardiomyopathy.

An important aspect of the clinical expression of mechanoelectric transduction is that pathological situations appear to amplify mechanoelectric feedback. This has been found when stretching several types of experimental preparations,[66,107-114] or under osmotic stress.[137] In addition, its interaction with the autonomic nervous system could be important, for autonomic derangements feature highly in cardiac pathology.

Mechanoelectric Transduction/Feedback as a Fundamental Surrogate in Arhythmia?

In order to support the contention that there is a fundamental surrogate, several criteria have to be fulfilled, and these can be described with the aid of Figure 10.

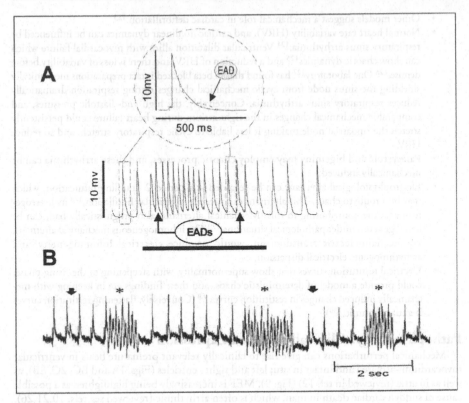

Figure 9. A) Four-second volume inflation (stretch) of intact isolated atrium produces EADs in the monophasic action potential, and a run of atrial tachycardia. On a faster time base (circle inset), compared with control (solid line) the stretch action potential (dashed line) shows the EAD as well as shortening in the plateaux phase (reprinted with permission from ref. 229). B) Records from intact atrium. Raising intra-atrial press (12.5 to 15 cm water) produces the occasional premature beat (vertical arrows), as well as episodes of nonsustained atrial fibrillation (reprinted with permission from ref. 230).

Mechanoelectric Feedback and Electrophysiological Arrhythmic Mechanisms

There are several electrophysiological mechanisms recognised to promote (initiate and sustain) arrhythmia. These include[138] single cell mechanisms such as afterdepolarisations and triggering (left hand side Fig. 10A), and multicellular mechanisms such as dispersion of repolarisation and reentry (right hand side Fig. 10A). If mechanoelectric feedback/transduction has any role in providing some mechanistic unification of the diverse clinical correlates, it should use these mechanisms in generating arrhythmia.

"Single Cell" Mechanisms

- Intracellular calcium changes play fundamentals roles in generating arrhythmia and calcium dynamics change in pathology.[139-145] Chronically, heart failure and hypertrophy are associated with abnormal calcium handling,[146-148] and arrhythmia. These disturbances favour afterdepolarisations. This section suggests that its no coincidence that Mechanoelectric feedback/transduction, roughly, changes intracellular calcium in comparable directions.[39] Calcium overload increases the likelihood of stretch arrhythmias.[149]
- Afterdepolarisations feature in generating arrhythmia,[138,150] and mechanically-induced afterdepolarisations appear able to reach threshold to initiate premature beats.[37,60] (This "hump" on the action potential has been described as an early afterdepolarisation, but a

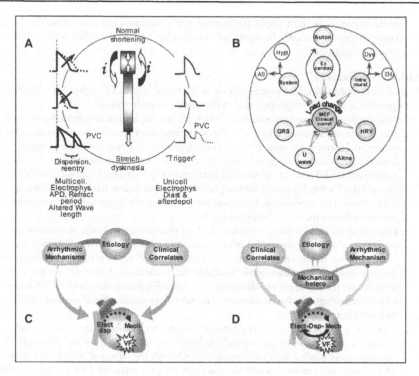

Figure 10. Commonalities in mechanoelectric transduction/feedback and pathology. A) Generalized electrophysiological arrhythmic mechanisms invoking mechanoelectric heterogeneity. Two myocardial segments diagrammed in the center. The normal segment (top) contracts normally if loaded with a normal segment in series. It produces a normal action potential (@ 2 o'clock). If it contracts against a weaker (say ischaemic) segment (bottom), it shortens more and produces a prolonged (dotted) action potential (superimposed @ 10 o'clock). In the case of the weaker stretched segment, the action potential (and refractory period) abbreviates (dotted, @ 9 o'clock). The stretched segment may also produce an after depolarisation (@ 3 o'clock), and premature ventricular beats (PVC). The repolarisation times are dispersed and produce current flow (curved arrows and (i) between the segments. Dispersion also depicted in the superimposed action potentials (@ 8 o'clock). Dispersion of refractory periods will promote changes in wave length and reentry (bottom left), with attendant PVC. B) Mechanoelectric feedback (MEF center) can relate to known clinical correlates of sudden death The load change can be (i) systemic (@10 o'clock) such as hypertension (Hypt) and aortic stenosis (AS); (ii) intramural (@ 2 o'clock) such as the dilated (Dil) chambers of ventricular cardiomyopathy or atrial dilation with mitral valve pathology. (iii) Extracorporeal (@12 o'clock: Ex cardiac) such as "Commotio Cordis". The autonomic nervous system (Auton: North of 12 o'clock) also interacts with abnormal load changes. Electrocardiographic correlates also exist (see text) with MEF: the signals themselves (QRS complexes: @ 8 o'clock, and u waves: @ 7 o'clock), and measures derived from the electrocardiogram (alternans: @ 5 o'clock, and heart rate variability: @ 4 o'clock). C) Conforming situation: eteology, arrhythmic mechanisms, and clinical correlates. The aeteological factors (top centre) produce (curved arrows) both the arrhythmic mechanisms and the clinical correlates, which produce, for example, electrical dispersion (Elect dsp) in the diagrammatic heart (Electrical changes can compromise (dotted arrow) mechanical function: "Mech"). The electrical dispersion is conducive to ventricular fibrillation (VF). D) The (novel) MEF contribution to arrhythmia, using arrhythmic mechanisms and clinical correlates. The eteological factors produce (downward connection) mechanical heterogeneity: "hetero" (see Fig. 3A). This produces (looped arrows) the arrhythmic mechanisms (see Fig. 3A) as well as the clinical correlates (see Fig. 3B). The resultant mechanoelectric feedback ("Mech" first) produces mechanoelectric dispersion, which now forms part of an interactive mechanoelectric feedback loop (solid and dotted arrows). Reprinted with permission from Elsevier.[228]

more appropriate term might be "mechanically activated depolarisation" or "stretch induced depolarisation", because their mechanisms probably differ from other early afterdepolarisations).

"Multicell Mechanisms"

Spatial electrophysiological heterogeneity is a crucial aspect of arrhythmic pathology, and can provide initiation, triggering, and maintenance of arrhythmia.

- Abnormal current flow between areas of myocardium, can initiate abnormal depolarisations, particularly in pathological hearts.[138] Sustained load shortens effective refractory period of the ventricles and at the same time increases dispersion of action potential duration.[19] The T-wave of the electrocardiogram is some function of electrophysiological inhomogeneity of repolarisation (different action potential durations, or QT intervals), and mechanical loading changes the T wave.[33] In many cardiac pathologies some regions have more stress and strain than others. This mechanical dispersion would favour the dispersion of repolarisation in the ventricles via mechanoelectric feedback.
- Changes in excitability facilitate reentry and are thus pro-arrhythmic. Mechanical changes can induce changes in myocardial refractoriness and excitability.[65] Not all studies, however, are convincing, but this maybe related to where the recordings were taken - heterogeneity of the expression of mechanoelectric feedback. Acute dilation of the left or right ventricle shortens the refractoriness and decreases ventricular fibrillation threshold.[151] With chronic dilatation, analogously, hearts are more vulnerable to ventricular fibrillation and fibrillation threshold is decreased.[19]
- Changes in conduction velocity promote reentry mechanism in arrhythmia. Although analogous changes can be mechanically induced, the results here are not clear-cut.[19] This particular issue needs a little more resolution. Overall, the effects of stretch on ventricular refractoriness and conduction velocity tend to decrease the myocardial wavelength, favouring inducibility of ventricular arrhythmias and decreasing fibrillation threshold.

Mechanoelectric Feedback/Transduction and Clinical Correlates of Arrhythmia

The following section addresses possible links between previously suggested clinical correlates with respect to arrhythmia, and mechanoelectric factors with respect to arrhythmia. The contribution could be through any or all of the electrophysiological mechanisms mentioned above. Some of the correlates appear obvious: clearly, a mechanical measure would provide the most likely candidate to support an arrhythmic correlation with mechanoelectric feedback. This would include load changes systemically, intramyocardially, and extracorporeally. Others correlations are less apparent, but nonetheless real.

Mechanical Correlates

- Dilated hearts, with poor ejection (i.e., heart failure) correlate with sudden death which is probably arrhythmic in about half the patients. (Although there has been some argument that not all cardiac-related sudden death is arrhythmic). Premature ventricular beats do not predict sudden death, and a powerful predictor of sudden death remains the left ventricular ejection fraction in heart failure.[152-154] Perhaps significantly, most antiarrhythmic drugs control arrhythmia poorly in heart failure, some even increasing sudden deaths.[155] This raises questions as to the validity of the therapeutic target when using pharmacological interventions.
- Wall motion disturbances (dyskinesia - Fig. 8B, and depicted in Figs. 10A and 10B @ 2 o'clock) are striking in regional ischaemia, and sudden arrhythmic death correlates with wall motion abnormalities.[156,157] In a tangential study[158] rapid pacing (implantable pacemaker) to control ventricular tachycardia reduced mechanoelectric feedback by curtailing abnormal wall motion. The latter was enhanced by long pauses, and this often preceded the arrhythmia. Global mechanical factors all involve changes in wall mechanics,

but importantly because of remodelling and patchy processes, regional heterogeneity of mechanical function is prominent.

- Systemic load changes (Fig. 10B @ 11 o'clock), experimentally, or in clinical situations such as hypertension and aortic stenosis, have been associated with arrhythmia (see ref. 9 for review).
- Acute extracorporeal mechanical load change (Fig. 10B - bottom circle @ 12 o'clock) by a sharp precordial blow can be associated with sudden death (*"Commotio cordis"*)[17,22,159] There is no postmortem evidence of contusion trauma to the heart, and death is more than likely a mechanically induced arrhythmia.

Electrophysiological Correlates

- An abnormal U wave in the ECG (Fig. 10B @ 7 o'clock) may be associated with arrhythmia,[160] particularly in long QT and torsade de pointes.[160,161] (And also hypokalaemia, see ref. 13.) The wave may be associated with inhomogeneous repolarisation late in the cardiac cycle, but it has been suggested that mechanical changes can generate U waves.[162,163] U wave changes in the epicardial ECG during load manipulation in vivo have been reported.[37]
- QRS (Fig. 10B @ 8 o'clock) complex changes reflect conduction defects, and the risk of arrhythmia was high when marked right ventricular enlargement was found together with QRS prolongation.[164] After repair of tetralogy of Fallot, chronic right ventricular volume overload, diastolic function, and QRS prolongation all correlated, with the latter an extremely sensitive predictor of potentially lethal ventricular arrhythmia.
- A reduction in heart rate variability (Fig. 10B @ 4 o'clock) is associated with ventricular dilatation, and myocardial failure (see also ref. 16).[126] Even in the absence of heart failure at the early stage of left ventricular dilatation, splinting the sinoatrial node reduces heart rate variability.[127] This prevents dynamic changes with respiration, reducing the high frequency component of the variability. Possibly near terminal situations produce dilated ventricles with high end-diastolic pressures spilling over to the right atrium. This would tense the sinoatrial node and reduce the high frequency component.
- Electrophysiological alternans (Fig. 10B @ 5 o'clock), in experimental ischaemia as well as in the clinical situation, can precede ventricular fibrillation, and the literature in this area is escalating.[128] A ventricular premature beat can either turn the electrical alternans on or off,[165] and in an analogous fashion, a sudden load change can turn a mechanical alternans off or on.[129] Cardiac failure and ischemia (which also show load changes) can modulate or produce electromechanical alternans.[166-171] Importantly, both electrical alternans and mechanical alternans,[172] can be heterogeneous. In the latter, mechanoelectric feedback would be heterogeneous, and pro-arrhythmic.

An interesting tangent is related to changes in the ST segment of the ECG during regional myocardial ischaemia. These changes are legion and are described in any respectable textbook on electrocardiography, with emphasis that they are directly related to the metabolic and ionic consequences of ischaemia. However one study has produced evidence that the ST changes are somehow a product of mechanical, and regional, bulging of the affected area.[173] The authors produced local bulging by regional blood perfusion containing negative inotropic agents.

Correlates Invoking Neuro-Humoral Components (Fig. 10B - Top Circle @ 12 O'Clock)

β Adrenoreceptor Blockade (β Blockade) Fig. 6 @ 11 O'Clock

One of the few therapeutic agents reducing mortality from sudden (arrhythmic) death in many subsets of patients and in related experimental situations are the β adrenergic receptor blockers. Study in this area is already voluminous, rising exponentially and a few recent reviews and studies are included here in refs. 174-181. The clinical situations covered include, infarction, hypertension, subaortic stenosis, cardiomyopathy, mitral valve prolapse, congestive cardiac failure. Moreover, depleting endogenous catecholamines by bretilium tosylate is one method for treating clinical arrhythmia.[182-187]

Although the mechanism of the therapeutic efficacy may well be due to a combination of direct anti-ischaemic effects and preservation of vagal tone, the above observations have elements of MMC. First, many of the clinical conditions associated with the autonomic nervous system and arrhythmia are associated with abnormal mechanical loads or wall motion. Secondly, the clinically related observations are in keeping with the mechanoelectric studies illustrated above (section I,d). That is, electrophysiology can be modified by autonomic agonists/antagonists. Moreover, bretylium tosylate, analogous to its therapeutic effects in abating clinical arrhythmia, curtails stretch arrhythmia.[82] We need a note of caution, for the action of bretylium may be directly electrophysiological.[188]

Following the heterogeneity theme, cardiac sympathetic enervation is heterogeneous. This point seems to be important because a reduced sympathetic cardiac enervation has been identified as a strong predictor of cardiac mortality in idiopathic dilated cardiomyopathy.[189] A heterogeneous enervation[189,190] would exacerbate any heterogeneous mechanoelectric coupling.

Renin-Angiotensin System

It is likely that the Renin Angiotensin system displays cross-talk with mechanotransduction, either by its action as a load-reducing peripheral vasodilator, or at the cell signal transduction level[191]). Either crosstalk mechanism could affect arrhythmia. Clinical trials have suggested that the reduction in sudden arrhythmic death observed with angiotensin converting enzyme (ACE) inhibitors may be related to its load reducing actions on electrophysiology (see reviews in this area in refs. 9,13,192). That is, by its action in reducing wall stress/strain.[193] However, cell signal crosstalk can also explain this effect, and this could be related to ACE inhibitor's slowing of cardiac remodelling. Cardiovascular drugs, such as the ACE inhibitors, and carvedilol β blocker activity which interact with the remodelling of heart have proved their efficacy in terms of cardiovascular mortality, antiarrhythmic action and prevention of sudden death.

Correlates Invoking Electrolyte Disturbances (Hypokalaemia)

MMC and Hypokalaemia

Diuretic therapy in patients can reduce serum potassium, which can be arrhythmogenic,[194-196] although interaction with other factors may be important.[197] Electrolyte depletion has been considered a risk factor.[198] Hypokalaemia can be related to experimental mechanotransduction.[100,199,200] For example, arrhythmia was generated in isolated heart, normally perfused, by increasing ventricular load. Perfusion with low potassium solutions increased the incidence/severity of this mechanically induced arrhythmia. A combination of Torsade de Point, hypokalaemia, and afterdeolarisations have been electrophysiologically implicated in arrhythmia,[160,161] and the latter two have been mechanically implicated.

Localised potassium increases during acute regional ischaemia,[201,202] which also produces dyskinesia (the affected area bulges, stretches, during systole).[203] Pilot experiments show that the early regional potassium rise after experimental coronary occlusion can be markedly reduced if the dyskinetic stretch is extraneously and mechanically prevented or restrained.[203a] That is, there could be a novel explanation for the potassium rise in acute regional ischaemia in that stretch opens channels to allow potassium to leak out of cells - perhaps via the KATP channel, which is mechanosensitive.

Remodelling and Cardiac Failure: Long Term Mechanoelectric Changes

Chronic increase in load produces cascades leading to hypertrophy and remodelling. Intracellular calcium, cell signals and the cytoskeleton would be possible contenders in the cascades which inevitably experience crosstalk (see for example, see ref. 204), switching on immediate early genes.

Myocardial failure with a dilated, stretched heart, is associated with sudden arrhythmic death. In keeping with the altered interactive gains in the pathological situation alluded to above, studies at the membrane level show, strikingly, that SACs are persistently activated in heart failure.[137] Moreover, mechanosensors are thought to be responsible for the downstream molecular changes[98] in remodelling ventricular structure and function in heart failure. This remodelling involves a crosstalk between several cellular mechanisms, which may also be conducive to arrhythmia, as early onset genes are turned on in parallel with changes in electrophysiology.[205] It appears that in the longer term, load-induced remodelling in global myocardial failure can produce subtler but significant mechanical heterogeneity, crosstalk, and thus electrophysiological heterogeneity, which is arrhythmogenic.

MMC and Ischaemic Preconditioning

Ischaemic preconditioning is a consequence of a short period of ischaemia protecting the myocardium from the damage produced by a subsequent more prolonged period of ischaemia. Stretch also seems to induce preconditioning.[206] This may be related to activation of adenosine receptors, (KATP) channels, and/or PKC. In one study, ischemic preconditioning was induced by 5 min ischaemia, and 5 minutes reperfusion. A transient volume overload was used for mechanical preconditioning. The infarct size-reducing effect of stretch, with no ischemia, was prevented by the stretch channel blocker Gd3+, and Glibencamide. This means that activation of mechanosensitive ion channels may produce an analogue of ischaemic preconditioning - probably by downstream activation of PKC, adenosine receptors, and/or KATP channels.[207]

Therapeutic Aspects

- Extracorporeal mechanical intervention: Mechanical precordial cardioversion, and "cough cardioversion" are recognised.[208] In both cases there is mechanical transmission to the myocardium. Thump-pacing (Fig. 11A) has been used in emergencies until pacemaker facilities are available. Conversely, increasing cardiac chamber volume increases fibrillation threshold,[19] a finding supported by mathematical modelling.[123]
- Intravascular mechanical intervention: One study has demonstrated a facilitatory effect of mechanoelectric feedback in controlling pharmacologically refractory arrhythmia. Mechanically lowering systolic afterload by intra-aortic balloon counterpulsation reduced the arrhythmia, and then enabled pharmacological regimes.[209] In an analogous fashion reducing preload can facilitate defibrillation.[210]
- Pharmacological: If stretch activated channels do contribute to stretch arrhythmia, and stretch arrhythmia to clinical arrhythmia, a stretch activated channel blocker may potentially provide a novel class of antiarrhythmic tool. Gadolinium, often used experimentally as a SAC blocker can curtail stretch arrhythmia in ventricle (Fig. 11B), is unsatisfactory because of its toxicity: perhaps similarly streptomycin, which can reduce mechanically induced electrical changes[70,72] which can also curtail stretch arrhythmia in atrium.[26,211] The search for a specific blocker is progressing, and one possibility is a tarantula venom, which appears able to curtail fibrillation in ventricle as well as atrium.[212] As the KATP channel is stretch sensitive, and it may be involved in the mechanically-induced arrhythmia of *"Commotio Cordis"*.[159] Blockers of this channel may also have potential.

Peripheral vasodilators which would reduce wall stress/strain, have a favourable effect on arrhythmia and this applies particularly with angiotensin converting enzyme (ACE) inhibitors.[9,19] ACE inhibitors could also operate intramyocardially by reducing the remodelling process, so reducing patchy dyskinesia. However, ACE inhibitors also have autonomic effects[213] and which could be favourable by increasing the baroreflex sensitivity:[214,215] both contribute to prevent ventricular arrhythmias.

One of the few therapeutic agents reducing sudden (arrhythmic) deaths in many subsets of patients and in related experimental situations are the β adrenergic blockers,[180,216-221] and these blockers can curtail stretch arrhythmia (see above). Moreover, depleting endogenous catecholamines by bretilium tosylate is clinically beneficial,[130,184] and may well do the same in stretch arrhythmia. Indirectly, thus, it could be that the β blockers contribute to their effects on mortality via their effects on mechanoelectric transduction.

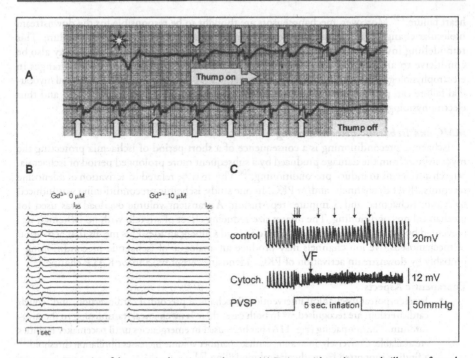

Figure 11. Examples of therapeutic directions and targets (A) Patient with cardiac standstill, apart from the occasional ectopic beat (asterisk). An external mechanical "thumper" (vertical block arrows) applied precordially (thump on), elicits ventricular beats (reprinted with permission from 231). B) Ventricular inflation about half a second after a beat at the beginning of a set of recordings in a control isolated heart (no gadolinium (Gd) regularly produces an ectopic beat. 10 micromolar Gadolinium suppresses these mechanically induced beats (reprinted with permission from ref. 61). C) Inflation of the ventricle to a peak ventricular systolic (PVSP) of 50 mmHg in an isolated heart under control conditions produces sporadic premature beats (vertical arrows in top trace). Depolymerising the cytoskeleton using cytochalasin (cytoch) "sensitises" the ventricle, and the same mechanical manoeuvre produces ventricular fibrillation (VF) (modified from ref. 223).

The cytoskeleton plays a crucial role in mechanotransduction, and it seems to feature in cardiac pathology. About half the deaths in cardiac failure are arrhythmic, and the cytoskeleton can be severely altered in this condition,[132,133] as well as in ischaemia.[222] A pilot study has revealed that depolymerisation of the cytoskeleton facilitates mechanically-induced fibrillation (Fig. 11C from ref. 223). An interesting question would be whether the converse, an analogue of polymerisation, quells stretch arrhythmias. We have to wait for more concrete evidence that the cytoskeleton could provide a therapeutic target.

Summary and Conclusions

William of Occam (1280-1349) was an English theologian and philosopher (the prevailing government and church regarded him as subversive.) He stated: *"Entia non sunt multiplicanda praeter necessitatum"*. This is otherwise known as Occam's Razor. Roughly interpreted as far as science is concerned, it proposes that rather than five hypotheses to explain (say) five observations, it is preferable to invoke one hypothesis to explain all of them. Cardiovascular clinical correlates from the perspective of predicting sudden arrhythmic death, run the gamut (depicted in Fig. 10B) from the fairly abstract, such as indices employing nonlinear dynamics ("chaos") and heart rate variability, through to electrophysiological phenomena such as late potentials, QT dispersion, and alternans; then on to mechanical indices like ejection fraction.

They even include psychosocial factors, which may be related to autonomic imbalance. Although some of these aspects have been related to mechanical factors in previous reviews (see refs. 9,13,14,16,162,224,225) this summary reassembles the foregoing text and presents a different slant. It bears on the possibility that many of the mechanical and arrhythmic factors impinge on a common pathway, and suggests that abnormalities in cardiac load provide some putative common path in the cascades of the clinical and monitoring observations. Boldly posed, it asks whether mechanoelectric coupling, transduction, or feedback, can be an embracing clinical correlate. Does mechanical heterogeneity provide one of the common final paths (Fig. 10D)?

In order to support the contention, several criteria have been fulfilled, and these can be described with the aid of Figure 10C and D. The curved arrows in Figure 10C, present the conventional dogma. The aeteological factors (Fig. 10C centre) through a variety of mechanisms, produce the various clinical correlations (right curved arrow) as well as the various electrophysiological arrhythmic mechanisms (left curved arrow). The clinical correlates or factors act via the arrhythmic mechanisms, and/or the arrhythmic mechanisms produce the clinical correlates and so precipitate lethal arrhythmia (VF) through, say, electrical dispersion. The argument being developed is presented in Figure 10D. The aeteological factors produce mechanical changes (downward block arrow) and these are common factors linking the clinical correlates with arrhythmic mechanisms (the 2 upwardly looped arrows) invoking dispersion of electromechanical interaction in the diagrammatic heart. This common link produces lethal arrhythmia (VF) via mechanoelectric dispersion.

The argument thus falls and develops within two broad categories which link together. First, the diverse factors contributing to, or correlating with arrhythmia impinge on, or interact with mechanoelectric mechanisms, including dispersion (Fig. 10C inside diagrammatic heart) - at the cellular as well as the gross level. There must be some sort of crosstalk.[14,16] Second, as there are several accepted basic cellular electrophysiological mechanisms precipitating or maintaining arrhythmia, mechanoelectric feedback uses most of these mechanisms.

References

1. Bainbridge FA. The influence of venous filling upon the rate of the heart. J Physiol 1915; 50:65-84.
2. Bozler E. Tonus changes in cardiac muscle and their significance for the initiation of impulses. Am J Physiol 1943; 139:477-80.
3. Penefsky ZJ, Hoffman BF. Effects of stretch on mechanical and electrical properties of cardiac muscle. Am J Physiol 1963; 204:433-8.
4. Deck KA. Anderungen des ruhe potentials and der kabeleigenschaflen von Purkinje-Faden bei der dehung. Pflugers Arch 1964; 280:131-40.
5. Dudel J, Trautwein W. Das aktionspotential und mechanogram des herzmuskels unter dem einflus der dehnung. Cardiologie 1954; 362.
6. Kaufmann R, Theophile U. Automatie-fordernde dehnungseffekte an purkinje-faden, papillarmuskeln und vorhoftrabekeln von rhesus-affen. Pflugers Arch 1967; 297:174-89.
7. Stauch-Von M. Electromechonische beziehungen am isolierten Frosch Herzon das monophasiesche aktionspotential bei isotoniescher und isometriescher kontraktion. Arch Kreisl Forch 1966; 49:1-4.
8. Lab MJ. Is there mechano-electric transduction in cardiac muscle? The monophasic action potential of the frog ventricle during isometric and isotonic contraction with calcium deficient perfusions. S Afr J Med Sci 1968; 33:60.
9. Dean JW, Lab MJ. Arrhythmia in heart failure: Role of mechanically induced changes in electrophysiology. Lancet 1989; 1:1309-12.
10. Lab MJ, Dean J. Myocardial mechanics and arrhythmia. J Cardiovasc Pharmacol 1991; 18(Suppl 2):S72-S79.
11. Sideris DA. High blood pressure and ventricular arrhythmias. Eur Heart J 1993; 14:1548-53.
12. Nazir SA, Lab MJ. Mechanoelectric feedback and atrial arrhythmias. Cardiovasc Res 1996; 32:52-61.
13. Reiter MJ. Effects of mechano-electric feedback: Potential arrhythmogenic influence in patients with congestive heart failure. Cardiovascular Res 1996; 32:44-51.
14. Lab MJ. Mechanoelectric feedback (transduction) in heart: Concepts and implications. Cardiovascular Res 1996; 32:3-14.
15. Taggart P, Sutton PMI. Cardiac mechano-electric feedback in man: Clinical relevance. Prog Biophys Mol Biol 1999; 71:139-154.

16. Lab MJ. Mechanosensitivity as an integrative system in the heart: An audit. Prog Biophysics and Molec Biol 1999; 71:7-27.
17. Kohl P, Hunter P, Noble D. Stretch-induced changes in heart rate and rhythm: Clinical observations, experiments and mathematical models. Prog Biophys Mol Biol 1999; 71:91-138.
18. Franz MR. Mechano-electrical feedback. Cardiovasc Res 2000; 45:263-6.
19. Reiter MJ. Contraction excitation feedback. In: Zipes DP, Jalife J, eds. Cardiac Electrophysiology from Cell to Bedside. Philadelphia: WB Saunders Company, 2000:249-55.
20. Tavi P, Laine M, Weckström M et al. Cardiac mechanotransduction: From sensing to disease and treatment. Trends Pharmacol Sci 2001; 22:254-60.
21. Guadalajara Boo JF. Mechanoelectric feedback and sudden death in heart failure. Arch Cardiol Mex 2001; 71(Suppl 1):S69-S75.
22. Babuty D, Lab MJ. Mechanoelectric contributions to sudden cardiac death. Cardiovasc Res 2001; 50:270-9.
23. Kamkin AG, Kiseleva IS, Iarygin VN. Ion mechanisms of the mechanoelectrical feedback in myocardial cells. Usp Fiziol Nauk 2001; 32:58-87.
24. Eckardt L, Kirchhof P, Breithardt G et al. Load-induced changes in repolarization: Evidence from experimental and clinical data. Basic Res Cardiol 2001; 96:369-80.
25. Hamill OP, Martinac B. Molecular basis of mechanotransduction in living cells. Physiol Rev 2001; 81:685-740.
26. Babuty D, Lab M. Heterogeneous changes of monophasic action potential induced by sustained stretch in atrium. J Cardiovasc Electrophysiol 2001; 12:323-9.
27. Kamkin A, Kiseleva I, Wagner KD et al. Mechanically induced potentials in fibroblasts from human right atrium. Exp Physiol 1999; 84:347-56.
28. Cooper PJ, Lei M, Cheng LX et al. Selected contribution: Axial stretch increases spontaneous pacemaker activity in rabbit isolated sinoatrial node cells. J Appl Physiol 2000; 89:2099-104.
29. James TN. The sinus node as a servomechanism. Circ Res 1973; 32:307-13.
30. Klein LS, Miles WM, Zipes DP. Effect of atrioventricular interval during pacing or reciprocating tachycardia on atrial size, pressure, and refractory period. Contraction-excitation feedback in human atrium. Circ 1990; 82:60-8.
31. Kaseda S, Zipes DP. Contraction-excitation feedback in the atria: A cause of changes in refractoriness. J Am Coll Cardiol 1988; 11:1327-36.
32. Zhang YH, Hancox JC. Gadolinium inhibits Na(+)-Ca(2+) exchanger current in guinea-pig isolated ventricular myocytes. Br J Pharmacol 2000; 130:485-8.
33. Dean JW, Lab MJ. Effect of changes in load on monophasic action potential and segment length of pig heart in situ. Cardiovasc Res 1989; 23:887-96.
34. Takagi S, Miyazaki T, Moritani K et al. Gadolinium suppresses stretch-induced increases in the differences in epicardial and endocardial monophasic action potential durations and ventricular arrhythmias in dogs. Jpn Circ J 1999; 63:296-302.
35. Brody DA. A theoretical analysis of intracavity blood mass on the heart lead relationship. Circ Res 1956; 4:731-8.
36. Choo MC, Gibson DG. U-Waves in ventricular hypertrophy: Possible demonstration of mechano-electric feedback. Br Heart J 1986; 55:428-33. ·
37. Franz MR, Burkhoff D, Yue DT et al. Mechanically induced action potential changes and arrhythmia in isolated and in situ canine hearts. Cardiovasc Res 1989; 23:213-23.
38. Monroe RG, Gamble WG, La-Farge CG et al. The Anrep effect reconsidered. J Clin Invest 1972; 2583.
39. Calaghan SC, White E. The role of calcium in the response of cardiac muscle to stretch. Prog Biophys Mol Biol 1999; 71:59-90.
40. Franz MR. Mechano-electrical feedback in ventricular myocardium. Cardiovasc Res 1996; 32:15-24.
41. Craelius W. Stretch-activation of rat cardiac myocytes. Exp Physiol 1993; 78:411-23.
42. La Farge CG, Monroe RG, Gamble WJ et al. Left ventricular pressure and norepinephrine efflux from the dennervated heart. Am J Physiol 1970; 219:519-24.
43. Kaufmann RL, Lab MJ, Hennekes R et al. Feedback interaction of mechanical and electrical events in the isolated mammalian ventricular myocardium (cat papillary muscle). Pflugers Arch 1971; 324:100-23.
44. Lab MJ, Allen DG, Orchard CH. The effects of shortening on myoplasmic calcium concentration and on the action potential in mammalian ventricular muscle. Circ Res 1984; 55:825-9.
45. Bremel RD, Weber A. Cooperation within actin filament in vertebrate skeletal muscle. Nat New Biol 1972; 238:97-101.
46. Zhang Y, Ter Keurs HE. Effects of gadolinium on twitch force and triggered propagated contractions in rat cardiac trabeculae. Cardiovasc Res 1996; 32:180-8.

47. Tavi P, Han C, Weckström M. Mechanisms of stretch-induced changes in [Ca2+]i in rat atrial myocytes: Role of increased troponin C affinity and stretch-activated ion channels. Circ Res 1998; 83:1165-77.

48. Gamble J, Taylor PB, Kenno KA. Myocardial stretch alters twitch characteristics and Ca2+ loading of sarcoplasmic reticulum in rat ventricular muscle. Cardiovasc Res 1992; 26:865-70.

49. Lab MJ, Zhou BY, Spencer CI et al. Length dependent changes of mechanical restitution in isolated superfused guinea pig papiallry muscle. J Physiol 1993; 459:506P.

50. Slinker BK, Shroff SG, Kirkpatrick RD et al. Left ventricular function depends on previous beat ejection but not previous beat pressure load. Circ Res 1991; 69:1051-7.

51. Guharay F, Sachs F. Stretch-activated single ion channel currents in tissue-cultured embryonic chick skeletal muscle. J Physiol 1984; 352:685-701.

52. Morris CE. Mechanosensitive ion channels. J Memb Biol 1990; 113:93-107.

53. French AS. Mechanotransduction. Annu Rev Physiol 1992; 54:135-52.

54. Sackin H. Mechanosensitive channels. Annu Rev Physiol 1995; 57:333-53.

55. Cazorla O, Pascarel C, Brette F et al. Modulation of ion channels and membrane receptor activity by stretch in cardiomyocytes. Possible mechanisms for mechanosensitivity. Prog Bioph Molec Biol 1999; 71:29-58.

56. Chepilko S, Zhou H, Sackin H et al. Permeation and gating properties of a cloned renal K+ channel. Am J Physiol 1995; 268:C389-401.

57. Craelius W, Chen V, el Sherif N. Stretch activated ion channels in ventricular myocytes. Biosci Rep 1988; 8:407-14.

58. Hu H, Sachs F. Stretch-activated ion channels in the heart. J Mol Cell Cardiol 1997; 29:1511-23.

59. Brady AJ. Mechanical properties of isolated cardiac myocytes. Physiol Rev 1991; 71:413-28.

60. Lab MJ. Mechanically dependent changes in action potentials recorded from the intact frog ventricle. Circ Res 1978; 42:519-28.

61. Hansen DE, Craig CS, Hondeghem LM. Stretch-induced arrhythmias in the isolated canine ventricle: Evidence for the importance of mechano-electrical feedback. Circ 1990; 81:1094-105.

62. Dick DJ, Harrison FG, O'Kane PD et al. "Preconditioning" of mechanically induced premature ventricular beats in the isolated rabbit heart. J Physiol 1993; 459:509P.

63. Franz MR, Burkhoff D, Yue DT. Mechano-electrical feedback in the intact isolated perfused canine heart. Circulation 1985; (Supp III):382.

64. Zabel M, Koller BS, Sachs F et al. Stretch-induced voltage changes in the isolated beating heart: Importance of the timing of stretch and implications for stretch-activated ion channels. Cardiovasc Res 1996; 32:120-30.

65. Dean JW, Lab MJ. Regional changes in myocardial refractoriness during load manipulation in the in-situ pig heart. J Physiol 1990; 429:387-400.

66. Horner SM, Lab MJ, Murphy CF et al. Mechanically induced changes in action potential duration and left ventricular segment length in acute regional ischaemia in the in situ porcine heart. Cardiovasc Res 1994; 28:528-34.

67. Horner SM, Murphy CF, Coen B et al. Sympathomimetic modulation of load-dependent changes in the action potential duration in the in situ porcine heart. Cardiovasc Res 1996; 32:148-57.

68. Horner SM, Dick DJ, Murphy CF et al. Cycle length dependence of the electrophysiological effects of increased load on the myocardium. Circ 1996; 94:1131-6.

69. Taggart P, Sutton P, Lab M et al. Effect of abrupt changes in ventricular loading on repolarization induced by transient aortic occlusion in humans. Am J Physiol 1992; 263:H816-23.

70. Gannier F, White E, Lacampagne A et al. Streptomycin reverses a large stretch induced increases in [Ca2+]i in isolated guinea pig ventricular myocytes. Cardiovasc Res 1994; 28:1193-8.

71. Salmon AH, Mays JL, Dalton GR et al. Effect of streptomycin on wall-stress-induced arrhythmias in the working rat heart. Cardiovasc Res 1997; 34:493-503.

72. Eckardt L, Kirchhof P, Monnig G et al. Modification of stretch-induced shortening of repolarization by streptomycin in the isolated rabbit heart. J Cardiovasc Pharmacol 2000; 36:711-21.

73. Baron A, van Bever L, Monnier D et al. A novel K(ATP) current in cultured neonatal rat atrial appendage cardiomyocytes. Circ Res 1999; 85:707-15.

74. Van Wagoner DR. Mechanosensitive gating of atrial ATP-sensitive potassium channels. Circ Res 1993; 72:973-83.

75. Matsuda N, Hagiwara N, Shoda M et al. Enhancement of the L-type Ca2+ current by mechanical stimulation in single rabbit cardiac myocytes. Circ Res 1996; 78:650-9.

76. Tabarean IV, Juranka P, Morris CE. Membrane stretch affects gating modes of a skeletal muscle sodium channel. Biophys J 1999; 77:758-74.

77. Maingret F, Patel AJ, Lesage F et al. Mechano- or acid stimulation, two interactive modes of activation of the TREK-1 potassium channel. J Biol Chem 1999; 274:26691-6.

78. Kim D. A mechanosensitive K+ channel in heart cells. Activation by arachidonic acid. J Gen Physiol 1992; 100:1021-40.
79. Perez NG, de Hurtado MC, Cingolani HE. Reverse mode of the Na+-Ca2+ exchange after myocardial stretch: Underlying mechanism of the slow force response. Circ Res 2001; 88:376-82.
80. Kawakubo T, Naruse K, Matsubara T et al. Characterization of a newly found stretch-activated KCa, ATP channel in cultured chick ventricular myocytes. Am J Physiol 1999; 276:H1827-H1838.
81. Lab MJ, Dick D, Harrison FG. Propranolol reduces stretch arrhythmia in isolated rabbit heart. J Physiol (Lond) 1992; 446:539P.
82. Dick DJ, Lab MJ, Harrison FG et al. A possible role of endogeneous catecholamines in stretch induced premature ventricular beats in the isolated rabbit heart. J Physiol 1994; 479:133P.
83. Drake-Holland AJ, Noble MI, Lab MJ. Acute pressure overload cardiac arrhythmias are dependent on the presence of myocardial tissue catecholamines. Heart 2001; 85:576.
84. Lerman BB, Engelstein ED, Burkhoff D. Mechanoelectrical feedback: Role of beta-adrenergic receptor activation in mediating load-dependent shortening of ventricular action potential and refractoriness. Circ 2001; 104:486-90.
85. Mark MD, Herlitze S. G-protein mediated gating of inward-rectifier K+ channels. Eur J Biochem 2000; 267:5830-6.
86. Ji S, John SA, Lu Y et al. Mechanosensitivity of the cardiac muscarinic potassium channel. A novel property conferred by Kir3.4 subunit. J Biol Chem 1998; 273:1324-8.
87. Pleumsamran A, Kim D. Membrane stretch augments the cardiac muscarinic K+ channel activity. J Membr Biol 1995; 148:287-97.
88. Wang N, Ingber DE. Probing transmembrane mechanical coupling and cytomechanics using magnetic twisting cytometry. Biochem Cell Biol 1995; 73:327-35.
89. Ingber DE. Tensegrity: The architectural basis of cellular mechanotransduction. Annu Rev Physiol 1997; 59:575-99.
90. Cantiello HF, Stow JL, Prat AG et al. Actin filaments regulate epithelial Na+ channel activity. Am J Physiol 1991; 261:C882-8.
91. Cantiello HF. Role of the actin cytoskeleton on epithelial Na+ channel regulation. Kidney Int 1995; 48:970-84.
92. Prat AG, Cantiello HF. Nuclear ion channel activity is regulated by actin filaments. Am J Physiol 1996; 270:C1532-43.
93. Gu Y, Gorelik J, Spohr HA et al. High-resolution scanning patch-clamp: New insights into cell function. FASEB J 2002; 16:748-50.
94. Sigurdson W, Ruknudin A, Sachs F. Calcium imaging of mechanically induced fluxes in tissue-Cultured chick heart: Role of stretch-activated ion channels. Am J Physiol 1992; 262:H1110-H1115.
95. Yamada KM, Geiger B. Molecular interactions in cell adhesion complexes. Curr Opin Cell Biol 1997; 9:76-85.
96. Gumbiner BM. Proteins associated with the cytoplasmic surface of adhesion molecules. Neuron 1993; 11:551-64.
97. Plopper GE, McNamee HP, Dike LE et al. Convergence of integrin and growth factor receptor signaling pathways within the focal adhesion complex. Mol Biol Cell 1995; 6:1349-65.
98. Sadoshima J, Izumo S. The cellular and molecular response of cardiac myocytes to mechanical stress. Annu Rev Physiol 1997; 59:551-71.
99. Dalton GR, Jones JV, Evans SJ et al. Wall stress-induced arrhythmias in the working rat heart as left ventricular hypertrophy regresses during captopril treatment. Cardiovasc Res 1997; 33:561-72.
100. Evans SJ, Levi AJ, Lee JA et al. EMD 57033 enhances arrhythmias associated with increased wall-Stress in the working rat heart. Clin Sci Colch 1995; 89:59-67.
101. Sideris DA, Pappas S, Siongas K et al. Effect of preload and afterload on ventricular arrhythmogenesis. J Electrocardiol 1995; 28:147-52.
102. Pathak CL. Effect of stretch on formation and conduction of electrical impulses in the isolated sinoauricular chamber of the frog's heart. Am J Physiol 1958; 192:111-3.
103. Allen DG, Kentish JC. The cellular basis of the length-tension relation in cardiac muscle. J Molec Cell Cariol 1985; 17:821-40.
104. Kentish JC, Wrzosek A. Changes in force and cytosolic Ca concentration after length changes in isolated rat ventricular trabeculae. J Physiol 1998; 506:431-44.
105. Le Guennec JY, White E, Gannier F et al. Stretch-induced increase in resting intracellular calcium concentration in single guinea-pig ventricular myocytes. Exp Physiol 1991; 76:975-8.
106. White E, Le Guennec JY, Nigretto JM et al. The effects of increasing cell length on auxotonic contractions; membrane potential and intracellular calcium transients in single guinea-pig ventricular myocytes. Exp Physiol 1993; 78:65-78.
107. Tobler HG, Gornic CC, Anderson RW et al. Electrophysiologic properties of the myocardial infarction border zone: Effects of transient aortic occlusion. Surgery 1986; 100:150-6.

108. Calkins H, Maughan WL, Kass DA et al. Electrophysiologic properties of the myocardial infarction border zone. Effects of transient aortic occlusion. Surgery 1979; 100:150-5.
109. Calkins H, Weisman HF, Levine JH et al. Effect of acute volume load on refractoriness and arrhythmia development in isolated chronically infarcted canine hearts. Circulation 1989; 79:687-97.
110. Zhou BY, Harrison FG, Dick DJ et al. Ventricular arrhythmogenesis is enhanced by mechanelectric feedback in regional ischaemic heart of the anaesthetised pig. J Physiol 1993; 473:184P.
111. Lab MJ, Zhou BY, Dick DJ et al. Stretch-induced ventricular fibrillation in normal and globally anoxic isolated guinea pig hearts. J Physiol 1993; 473:183P.
112. Wang Z, Taylor LK, Denney WD et al. Initiation of ventricular extrasystoles by myocardial stretch in chronically dilated and failing canine left ventricle. Cir 1994; 90:2022-31.
113. Pye MP, Cobbe SM. Arrhythmogenesis in experimental models of heart failure: The role of increased load. Cardiovasc Res 1996; 32:248-57.
114. Kamkin A, Kiseleva I, Isenberg G. Stretch-activated currents in ventricular myocytes: Amplitude and arrhythmogenic effects increase with hypertrophy. Cardiovasc Res 2000; 48:409-20.
115. Vetter FJ, McCulloch AD. Mechanoelectric feedback in a model of the passively inflated left ventricle. Ann Biomed Eng 2001; 29:414-26.
116. Press WH, Flannery BP, Teukolsky SA et al. Numerical recipes, the art of scientific computing. Numerical Recipes 1986; 11:335-77.
117. Knudsen Z, Holden AV, Brindley J. Qualitative modeling of mechanoelectrical feedback in a ventricular cell. Bull Math Biol 1997; 59:1155-81.
118. Riemer TL, Sobie EA, Tung L. Stretch-induced changes in arrhythmogenesis and excitability in experimentally based heart cell models. Am J Physiol 1998; 275:H431-42.
119. Han C, Tavi P, Weckström M. Modulation of action potential by $[Ca2+]i$ in modeled rat atrial and guinea pig ventricular myocytes. Am J Physiol Heart Circ Physiol 2002; 282:H1047-H1054.
120. Rice JJ, Winslow RL, Dekanski J et al. Model studies of the role of mechano-sensitive currents in the generation of cardiac arrhythmias. J Theoretical Biol 1998; 190:295-312.
121. Markhasin VS, Solovyova O, Katsnelson LB et al. Mechano-electric interactions in heterogeneous myocardium: development of fundamental experimental and theoretical models. Prog Biophys Mol Biol 2003; 82(1-3):207-20.
122. Goldberger AL, West BJ. Chaos in physiology: Health or disease? In: Degn H, Holden AV, Olsen LF, eds. Chaos in biological systems. New York: Plenum, 1987:1-4.
123. Trayanova N. Concepts of ventricular fibrillation. Phi Trans Roy Soc 2001; 359:1327-1337.
124. Fortrat JO, Yamamoto Y, Hughson RL. Respiratory influences on nonlinear dynamics of heart rate variability in humans. None Specified 1997.
125. Poon CS, Merrill CK. Decrease of cardiac chaos in congestive heart failure. Nature 1997; 389:492-5.
126. Fauchier L, Babuty D, Cosnay P et al. Prognostic value of heart rate variability for sudden death and major arrhythmic events in patients with idiopathic dilated cardiomyopathy. J Am Coll Cardiol 1999; 33:1203-7.
127. Horner SM, Murphy CF, Coen B et al. Contribution to heart rate variability by mechanoelectric feedback. Stretch of the sinoatrial node reduces heart rate variability. Circ 1996; 94:1762-7.
128. Armoundas AA, Tomaselli GF, Esperer HD. Pathophysiological basis and clinical application of T-wave alternans. J Am Coll Cardiol 2002; 40:207-17.
129. Murphy CF, Lab MJ, Horner SM et al. Load induced period doubling bifurcations in the anaesthetised pig heart. Chaos Solitons and Fractals 1995; 5:707-12.
130. Garfinkel A, Kim YH, Voroshilovsky O et al. Preventing ventricular fibrillation by flattening cardiac restitution. Proc Natl Acad Sci USA 2000; 97:6061-6.
131. Allessie MA, Boyden PA, Camm AJ et al. Pathophysiology and prevention of atrial fibrillation. Circ 2001; 103:769-77.
132. Schaper J, Froede R, Hein S et al. Impairment of the myocardial ultrastructure and changes of the cytoskeleton in dilated cardiomyopathy. Circ 1991; 83:504-14.
133. Hein S, Kostin S, Heling A et al. The role of the cytoskeleton in heart failure. Cardiovasc Res 2000; 45:273-8.
134. Luna EJ, Hitt AL. Cytoskeleton—plasma membrane interactions. Science 1992; 258:955-64.
135. Cantiello HF. Role of the actin cytoskeleton in the regulation of the cystic fibrosis transmembrane conductance regulator. Exp Physiol 1996; 81:505-14.
136. Cantiello HF. Actin filaments stimulate the Na(+)-K(+)-ATPase. Am J Physiol 1995; 269:F637-43.
137. Clemo HF, Stambler BS, Baumgarten CM. Swelling-activated chloride current is persistently activated in ventricular myocytes from dogs with tachycardia-induced congestive heart failure. Circ Res 1999; 84:157-65.
138. Janse MJ. Electrophysiology of arrhythmias. Arch Mal Coeur Vaiss 1999; 92(Spec No 1):9-16.
139. Opie LH. Calcium antagonists, ventricular arrhythmias, and sudden cardiac death: A major challenge for the future. J Cardiovasc Pharmacol 1991; 18(Suppl 10):S81-6.

140. Opie LH. Mechanisms whereby calcium channel antagonists may protect patients with coronary artery disease. Eur Heart J 1997; 18(Suppl A):A92-104.

141. Levy MN, Wiseman MN. Electrophysiologic mechanisms for ventricular arrhythmias in left ventricular dysfunction: Electrolytes, catecholamines and drugs. J Clin Pharmacol 1991; 31:1053-60.

142. Di Diego JM, Antzelevitch C. High [Ca2+]o-induced electrical heterogeneity and extrasystolic activity in isolated canine ventricular epicardium. Phase 2 reentry. Circ 1994; 89:1839-50.

143. Schouten VJ, Vliegen HW, van der LA et al. Altered calcium handling at normal contractility in hypertrophied rat heart. J Mol Cell Cardiol 1990; 22:987-98.

144. Balke CW, Shorofsky SR. Alterations in calcium handling in cardiac hypertrophy and heart failure. Cardiovasc Res 1998; 37:290-9.

145. Wickenden AD, Kaprielian R, Kassiri Z et al. The role of action potential prolongation and altered intracellular calcium handling in the pathogenesis of heart failure. Cardiovasc Res 1998; 37:312-23.

146. Del Monte F, Johnson CM, Stepanek AC et al. Defects in calcium control. J Card Fail 2002; 8:S421-S431.

147. Zaugg CE, Buser PT. When calcium turns arrhythmogenic: Intracellular calcium handling during the development of hypertrophy and heart failure. Croat Med J 2001; 42:24-32.

148. Houser SR, Piacentino IIIrd V, Weisser J. Abnormalities of calcium cycling in the hypertrophied and failing heart. J Mol Cell Cardiol 2000; 32:1595-607.

149. Ferrier GR. The effects of tension on acetylstrophanthidin-induced transient depolarizations and aftercontractions in canine myocardial and Purkinje tissues. Circ Res 1976; 38:156-62.

150. el Sherif N. Early afterdepolarizations and arrhythmogenesis. Experimental and clinical aspects. Arch Mal Coeur Vaiss 1991; 84:227-34.

151. Jalal S, Williams GR, Mann DE et al. Effect of acute ventricular dilatation on fibrillation thresholds in the isolated rabbit heart. Am J Physiol 1992; 263:H1306-H1310.

152. Kelly M, Thompson P, Quinlan M. Prognostic significace of left ventricular ejection fraction after acute myocardial infarction. Br Heart J 1985; 53:16-24.

153. Copie X, Hnatkova K, Blankoff I et al. Risk of mortality after myocardial infarction: Value of heart rate, its variability and left ventricular ejection fraction. Arch Mal Coeur Vaiss 1996; 89:865-71.

154. Odemuyiwa O, Malik M, Farrell T et al. Multifactorial prediction of arrhythmic events after myocardial infarction. Combination of heart rate variability and left ventricular ejection fraction with other variables. Pace - Pacing Clin Electrophysiol 1991; 14:1986-91.

155. Preliminary report: Effect of encainide and flecainide on mortality in a randomized trial of arrhythmia suppression after myocardial infarction. The Cardiac Arrhythmia Suppression Trial (CAST) Investigators. N Engl J Med 1989; 321:406-12.

156. Shen WF, Cui LQ, Wang MH et al. Spontaneous alterations in left ventricular regional wall motion after acute myocardial infarction. Chin Med J (Engl) 1990; 103:1015-8.

157. Siogas K, Pappas S, Graekas G et al. Segmental wall motion abnormalities alter vulnerability to ventricular ectopic beats associated with acute increases in aortic pressure in patients with underlying coronary artery disease. Heart 1998; 79:268-73.

158. Perticone F, Ceravolo R, Maio R et al. Mechano-electric feedback and ventricular arrhythmias in heart failure. The possible role of permanent cardiac stimulation in preventing ventricular tachycardia. Cardiologia 1993; 38:247-52.

159. Link MS, Wang PJ, VanderBrink BA et al. Selective activation of the K(+)(ATP) channel is a mechanism by which sudden death is produced by low-energy chest-wall impact (Commotio cordis). Circ 1999; 100:413-8.

160. Bonatti V, Finardi A, Botti G. Recording of monophasic action potentials of the right ventricle in a case of long QT and isolated alternation of the U wave. Arch Mal Coeur 1979; 72:1180-6.

161. Cranefield PF, Aronson. Torsades de pointes and early afterdepolarizations. Cardiovasc Drugs Ther 1991; 5:531-7.

162. Lab MJ. Contraction-excitation feedback in myocardium: Physiological basis and clinical relevance. Circ Res 1982; 50:757-66.

163. Surawicz B. U wave: Facts, hypotheses, misconceptions, and misnomers. J Cardiovasc Electrophysiol 1998; 9:1117-28.

164. Gatzoulis MA, Till JA, Somerville J et al. Mechanoelectrical interaction in tetralogy of Fallot. QRS prolongation relates to right ventricular size and predicts malignant ventricular arrhythmias and sudden death [see comments]. Circ 1995; 92:231-7.

165. Dilly SG, Lab MJ. Electrophysiological alternans and restitution during acute regional ischaemia in myocardium of anaesthetized pig. J Physiol (Lond) 1988; 402:315-33.

166. Eyer KM. U wave alternans: An electrocardiographic sign of left ventricular failure. Am Heart J 1974; 87:41-5.

167. Freeman AB, Steinbrook RA. Recurrence of pulsus alternans after fentanyl injection in a patient with aortic stenosis and congestive heart failure. Can Anaesth Soc J 1985; 32:654-7.
168. Hastings HM, Fenton FH, Evans SJ et al. Alternans and the onset of ventricular fibrillation. Phys Rev E Stat Phys Plasmas Fluids Relat Interdiscip Topics 2000; 62:4043-8.
169. Luomanmaki K, Heikkila J, Hartikainen M. T-wave alternans associated with heart failure and hypomagnesemia in alcoholic cardiomyopathy. Eur J Cardiol 1975; 3:167-70.
170. Narayan P, McCune SA, Robitaille PM et al. Mechanical alternans and the force-frequency relationship in failing rat hearts. J Mol Cell Cardiol 1995; 27:523-30.
171. Ryan JM, Scheive JF, Hull HB et al. Experiences with pulsus alternans; ventricular alternation and the stage of heart failure. Circ 1956; 14:1099-103.
172. Murphy CF, Lab MJ, Horner SM et al. Regional electromechanical alternans in anesthetized pig hearts: Modulation by mechanoelectric feedback. Am J Physiol 1994; 267:H1726-35.
173. Shimada M, Nakamura Y, Iwanaga S et al. Nonischemic ST-segment elevation induced by negative inotropic agents. Jpn Circ J 1999; 63:610-6.
174. Reiter MJ, Reiffel JA. Importance of beta blockade in the therapy of serious ventricular arrhythmias. Am J Cardiol 1998; 82:9I-19I.
175. Sager PT. Modulation of antiarrhythmic drug effects by beta-adrenergic sympathetic stimulation. Am J Cardiol 1998; 82:20I-30I.
176. Campbell RW. ACE inhibitors and arrhythmias. Heart 1996; 76:79-82.
177. Wiesfeld AC, Crijns HJ, Tuininga YS et al. Beta adrenergic blockade in the treatment of sustained ventricular tachycardia or ventricular fibrillation. Pacing Clin Electrophysiol 1996; 19:1026-35.
178. Van Gelder IC, Brugemann J, Crijns HJ. Current treatment recommendations in antiarrhythmic therapy. Drugs 1998; 55:331-46.
179. Singh BN. Antiarrhythmic drugs: A reorientation in light of recent developments in the control of disorders of rhythm. Am J Cardiol 1998; 81:3D-13D.
180. Hjalmarson A. Effects of beta blockade on sudden cardiac death during acute myocardial infarction and the postinfarction period. Am J Cardiol 1997; 80:35J-9J.
181. Kennedy HL. Beta blockade, ventricular arrhythmias, and sudden cardiac death. Am J Cardiol 1997; 80:29J-34J.
182. Kowey PR. An overview of antiarrhythmic drug management of electrical storm. Can J Cardiol 1996; 12(Suppl B):3B-8B.
183. von Planta M, Chamberlain D. Drug treatment of arrhythmias during cardiopulmonary resuscitation. A statement for the Advanced Life Support Working Party of the European Resuscitation Council. Resuscitation 1992; 24:227-32.
184. Waller DG. Treatment and prevention of ventricular fibrillation: Are there better agents? Resuscitation 1991; 22:159-66.
185. Chamberlain DA. Lignocaine and bretylium as adjuncts to electrical defibrillation. Resuscitation 1991; 22:153-7.
186. Anderson JL. Bretylium tosylate: Profile of the only available class III antiarrhythmic agent. Clin Ther 1985; 7:205-24.
187. Duff HJ, Roden DM, Yacobi A et al. Bretylium: Relations between plasma concentrations and pharmacologic actions in high-frequency ventricular arrhythmias. Am J Cardiol 1985; 55:395-401.
188. Orts A, Alcaraz C, Delaney KA et al. Bretylium tosylate and electrically induced cardiac arrhythmias during hypothermia in dogs. Am J Emerg Med 1992; 10:311-6.
189. Merlet P, Benvenuti C, Moyse D et al. Prognostic value of MIBG imaging in idiopathic dilated cardiomyopathy. J Nucl Med 1999; 40:917-23.
190. Hainsworth R. Reflexes from the heart. Physiol Rev 1991; 71:617-58.
191. Eriksson SV, Eneroth P, Kjekshus J et al. Neuroendocrine activation in relation to left ventricular function in chronic severe congestive heart failure: A subgroup analysis from the Cooperative North Scandinavian Enalapril Survival Study (CONSENSUS). Clin Cardiol 1994; 17:603-6.
192. Pye P, Cobbe SM. Mechanisms of ventricular arrhythmias in cardiac failure and hypertrophy. Cardiovascular Res 1992; 26:740-50.
193. Cohn JN, Archibald DG, Ziesche S et al. Effect of vasodilator therapy on mortality in chronic congestive heart failure. Results of a Veterans Administration Cooperative Study. N Engl J Med 1986; 314:1547-52.
194. Stewart DE, Ikram H, Espiner EA et al. Arrhythmogenic potential of diuretic induced hypokalaemia in patients with mild hypertension and ischaemic heart disease. Br Heart J 1985; 54:290-7.
195. Gulker H, Haverkamp W, Hindricks G. Ion regulation disorders and cardiac arrhythmia. The relevance of sodium, potassium, calcium, and magnesium. Arzneimittelforschung 1989; 39:130-4.
196. Podrid PJ. Potassium and ventricular arrhythmias. Am J Cardiol 1990; 65:33E-44E.

197. Dargie HJ, Cleland JG, Leckie BJ et al. Relation of arrhythmias and electrolyte abnormalities to survival in patients with severe chronic heart failure. Circ 1987; 75:IV98-I107.
198. Packer M, Gottlieb SS, Blum MA. Immediate and long-term pathophysiologic mechanisms underlying the genesis of sudden cardiac death in patients with congestive heart failure. Am J Med 1987; 82:4-10.
199. James MA, Jones JV. The paradoxical role of left ventricular hypertrophy in wall stress-related arrhythmia. J Hypertens 1992; 10:167-72.
200. Dick DJ, Lab MJ. Effect of manipulation of potassium concentration on stretch-induced arrhythmia in the isolated Langendorff rabbit heart. J Physiol 1995; 487:140P, (ref Type: Abstract).
201. Blake K, Clusin WT, Franz MR et al. Mechanism of depolarization in the ischaemic dog heart: Discrepancy between T-Q potentials and potassium accumulation. J Physiol Lond 1988; 397:307-30.
202. Curtis MJ. The rabbit dual coronary perfusion model: A new method for assessing the pathological relevance of individual products of the ischaemic milieu: Role of potassium in arrhythmogenesis. Cardiovasc Res 1991; 25:1010-22.
203. Kessler M, Klovekorn WP, Hoper J et al. Local oxygen supply and regional wall motion of the dog's heart during critical stenosis of the LAD. Adv Exp Med Biol 1984; 169:331-40.
203a. Lab MJ. Regional stretch effects in pathological myocardium. In: Kohl, Franz, Sachs, eds. Cardiac mechano-electric feedback and arrhythmias: from pipette to patient. Elsevier, 2005.
204. Bustamante JO, Ruknudin A, Sachs F. Stretch-activated channels in heart cells: Relevance to cardiac hypertrophy. J Cardiovasc Pharmacol 1991; 17(Suppl 2):S110-S113.
205. Meghji P, Nazir SA, Dick DJ et al. Regional workload induced changes in electrophysiology and immediate early gene expression in intact in situ porcine heart. J Mol Cell Cardiol 1997; 29:3147-55.
206. Ovize M, Kloner RA, Przyklenk K. Stretch preconditions canine myocardium. Am J Physiol 1994; 266:H137-46.
207. Gysembergh A, Margonari H, Loufoua J et al. Stretch-induced protection shares a common mechanism with ischemic preconditioning in rabbit heart. Am J Physiol 1998; 274:H955-H964.
208. Caldwell G, Millar G, Quinn E et al. Simple mechanical methods for cardioversion: Defence of the precordial thump. BMJ 1985; 291:627-30.
209. Fotopoulos GD, Mason MJ, Walker S et al. Stabilisation of medically refractory ventricular arrhythmia by intra-aortic balloon counterpulsation. Heart 1999; 82:96-100.
210. Strobel JS, Kay GN, Walcott GP et al. Defibrillation efficacy with endocardial electrodes is influenced by reductions in cardiac preload. J Interv Card Electrophysiol 1997; 1:95-102.
211. Nazir SA, Dick DJ, Lab MJ. Mechanoelectric feedback and arrhythmia in the atrium of the isolated, Langendorf-perfused guinea pig hearts and its modulation by Streptomycin. J Physiol 1994, (Ref Type: Abstract).
212. Bode F, Sachs F, Franz MR. Tarantula peptide inhibits atrial fibrillation. Nature 2001; 409:35-6.
213. Young JB. Angiotensin-converting enzyme inhibitors post-myocardial infarction. Cardiol Clin 1995; 13:379-90.
214. Kowey PR, Verrier RL, Lown B. Effects of prostacyclin (PGI2) on vulnerability to ventricular fibrillation in the normal and ischemic canine heart. Eur J Pharmacol 1982; 80:83-91.
215. Kowey PR, Verrier RL, Lown B. Decreased vulnerability to ventricular fibrillation by vasodilator-induced baroreceptor sensitisation. Cardiovasc Res 1983; 17:106-12.
216. Aronow WS. Postinfarction use of beta-blockers in elderly patients. Drugs Aging 1997; 11:424-32.
217. Gundersen T. Secondary prevention after myocardial infarction: Subgroup analysis of patients at risk in the Norwegian Timolol Multicenter Study. Clin Cardiol 1985; 8:253-65.
218. Haverkamp W, Gulker H, Hindricks G et al. Effects of beta-blockade on the incidence of ventricular tachyarrhythmias during acute myocardial ischemia: Experimental findings and clinical implications. Basic Res Cardiol 1990; 85(Suppl 1):293-303.
219. Kendall MJ, Lynch KP, Hjalmarson A et al. Beta-blockers and sudden cardiac death. Ann Intern Med 1995; 123:358-67.
220. Olsson G, Ryden L. Prevention of sudden death using beta-blockers. Review of possible contributory actions. Circulation 1991; 84:VI33-7.
221. Wikstrand J, Kendall M. The role of beta receptor blockade in preventing sudden death [see comments]. Eur Heart J 1992; 13(Suppl D):111-20.
222. Krams R, Janssen M, Van der Lee C et al. Loss of elastic recoil in postischemic myocardium induces rightward shift of the systolic pressurevolume relationship. Am J Physiol 1994; 267:H1557-64.
223. Dick DJ, Coen BA, Lab MJ. The effect of cytochalasin-B on stretch induced arrhythmia in the Langendorff perfused isolated rabbit heart. J Physiol 1995; 489:166P, (ref Type: Abstract).
224. Franz MR. Mechano-electrical feedback in ventricular myocardium. Cardiovasc Res 1996; 32:15-24.
225. Taggart P, Sutton PMI, Treasure T et al. Contraction excitation feedback in man. Br Heart J 1988; 59:109.

226. Zeng T, Bett GC, Sachs F. Stretch-activated whole cell currents in adult rat cardiac myocytes. Am J Physiol Heart Circ Physiol 2000; 278:H548-H557.
227. Greve G, Lab MJ, Chen R et al. Right ventricular distension alters monophasic action potential duration during pulmonary arterial occlusion in anaesthetised lambs: Evidence for arrhythmogenic right ventricular mechanoelectrical feedback. Exp Physiol 2001; 86:651-7.
228. Lab MJ. Mechanoelectric Transduction/Feedback: Prevalence and Pathophysiology. In: Zipes DP, Jalife J, eds. Philadelphia: WB Saunders Company, 2003.
229. Nazir SA, Lab MJ. Mechanoelectric feedback in the atrium of the isolated guinea- pig heart. Cardiovasc Res 1996; 32:112-9.
230. Bode F, Katchman A, Woosley RL et al. Gadolinium decreases stretch-induced vulnerability to atrial fibrillation. Circ 2000; 101:2200-5.
231. Zoll PM, Belgard BSEE, Weintraub MJ et al. External mechanical cardiac stimulation. N Engl J Med 1976; 294:1274-5.

CHAPTER 5

Mechanotransduction in Cardiac Remodeling and Heart Failure

Jeffrey H. Omens,* Andrew D. McCulloch and Ilka Lorenzen-Schmidt

Abstract

Mechanotransduction is the process by which the cells of the heart convert mechanical signals to chemical signals responsible for cellular adaptation and remodeling. When this system cannot meet the demands of increased loading conditions, the cellular response will not be adequate, and eventually the pumping function of the heart will fail. Mechanical signaling and force transmission within and outside the myocyte are important players in the mechanotransduction process, and the cytoskeleton is a key component in the structural link between the force-generating sarcomere, the cell membrane and putative intracellular stress-sensing components. Several defects in cytoskeletal components have been linked to cardiac dilation and heart failure. LIM proteins are one such structural component of the cytoskeleton, and defects in these proteins lead to both right and left ventricular dysfunction. Although these proteins may have chemical signaling roles in mechanotransduction, their structural role in force transmission and mechanical signaling is being investigated and characterized. Thus, there is evidence that structural components of the myocardium such as the myocyte cytoskeleton play a critical role in mechanotransduction and are part of the mechanism behind cardiac remodeling and eventual heart failure.

Introduction: Cardiac Remodeling and Failure Are Dependent on Mechanical Forces

As the population ages and survival of initial acute cardiac events increases, heart failure is becoming more and more prevalent, yet clinical treatments have not kept pace with the disease. Heart failure is a deficiency in the pumping capacity of the ventricles, and has a wide variety of causes including ischemia, valvular disease, hypertrophy and cardiomyopathy.[1] The changes seen in the myocardium leading to heart failure are usually described as a process of cardiac "remodeling". Ventricular dilation, fibrosis, apoptosis are all components of tissue remodeling, that can have a negative impact on the performance of the heart. Remodeling can also be a compensatory process in the heart by which the heart can accommodate abnormal conditions. Compensated hypertrophy is one such case, in which the cells and tissue have remodeled in a positive fashion to accommodate excessive loads placed on the pumping chambers. Since it is well known that that the hypertrophic response of myocytes is regulated by mechanical forces,[2] it is natural to postulate that heart failure may be caused by abnormalities in cardiac myocyte mechanical signaling pathways.

*Corresponding Author: Jeffrey H. Omens—University of California San Diego, Department of Medicine, 0613J, La Jolla, California 92093-0613, U.S.A. Email: jomens@ucsd.edu

Cardiac Mechanotransduction, edited by Matti Weckström and Pasi Tavi.
©2007 Landes Bioscience and Springer Science+Business Media.

The heart is a mechanical pump, and the main purpose of the contractile cells of the heart is to generate forces that are transmitted through the myocyte fiber architecture and extracellular matrix into the ventricular pressure that drives cardiac output. Mechanical signals and force transmission are key players in the normal functioning of myocytes. The actin-myosin complex generates forces in response to intracellular calcium transients, and these molecular forces must be transferred through the myofilament lattice and cytoskeleton to the cell membrane, and across the membrane to neighboring cells via membrane-spanning complexes and the extracellular matrix. Forces are also transmitted "outside-in", from the extracellular environment into the myocytes, where they can be sensed by the appropriate machinery and produce signals that the cells need to maintain their phenotype.[3] Thus, mechanical stress is a key signal for the maintenance of myocyte structure and function, and mechanotransduction is the process that interprets these stress signals. If the stress response cannot be appropriately processed, dysregulated adaptation and remodeling may occur and eventually lead to heart failure.

We suspect that mechanical forces (stresses) play a critical role in cardiac remodeling and heart failure, both in force transmission out of the myocyte and stress sensing (mechanotransduction) within the cell. In dilated cardiomyopathy and heart failure, stresses may be elevated and reach a threshold level, above which the heart is no longer able to use its compensatory hypertrophy and remodeling responses to halt the downward progression. Because mechanical forces are transmitted simultaneously from the myofilaments to neighboring cells, to the extracellular matrix, and finally to the ventricular chambers, and because specific force transmission pathways depend on the detailed three-dimensional cytoarchitecture and mechanical properties, understanding these mechanisms is inherently difficult. Since the cytoskeleton is thought to be the structural component inside the cell responsible for transmission of forces in both directions,[4,5] from the sarcomere to the outside and also from extracellular environment to the mechanotransducer within the cell, and is known to be an important player in heart failure,[6] this component of the structural system will be the specific focus of this chapter.

Myocyte Response to External Loads

Myocytes can sense a wide range of external stimuli and initiate a variety of internal responses. External mechanical loading of myocytes can increase protein synthesis rates and cause cellular hypertrophy.[7-9] But less is known about the mechano-chemical transduction processes by which this external stimulus initiates the nuclear responses necessary for altered gene expression and eventual growth of the cells.[10,11]

When external loads are chronically elevated, adaptive cardiac hypertrophy develops to accommodate the increased load. Although certain forms of hypertrophy such as those induced by exercise training can be useful physiologic adaptations,[8] hypertrophy is considered a pathologic process when the muscle becomes structurally abnormal and functionally impaired. In volume overload or eccentric hypertrophy, myocytes respond to an increase in diastolic pressure and volume with series replication of sarcomeres, increased axial myocyte length and chamber dilation.[12,13] However, the consequent increase in systolic wall stress can also lead to wall thickening so that the ratio of ventricular radius to wall thickness changes little in this mode of hypertrophy. This is in contrast to the increased myocyte cross-sectional area by the parallel addition of myofibrils[14] and substantial wall thickening (concentric hypertrophy) during pressure overload.[12,15] Thus, the most prominent end-product of mechanotransduction in the myocyte is growth of the individual cell, and most remodeling processes in the heart involve either normal or abnormal myocyte hypertrophy.

The response to mechanical stimulation has been investigated directly with cultured muscle cell preparations, and many factors have been found to modulate the stretch response. Stretch-induced adaptation of cultured neonatal cardiocytes has been investigated with cells grown on elastic substrates.[9,16] Simpson and coworkers[17] have shown that mechanical loading is an important factor in the maintenance of myofibrillar structure in cultured cells. This loading may be related to internal loading due to phasic myocyte contraction or external loads

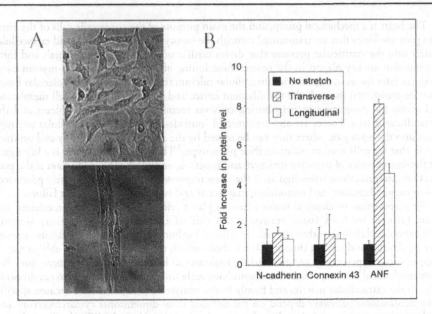

Figure 1. Micropatterning of neonatal cardiac myocytes. A) Patterned and control cells. Myocytes were cultured on a deformable membrane with a collagen substrate patterned with 10 micron wide stripes. The myocytes align typically either 1 or 2 cells wide on this type of pattern, with an adult-like phenotype. On control membranes (continuous collagen layer), cells take on a random, stellate shape. B) Protein concentrations after stretching cells. Cell were stretched anisotropically for 24 hours with either 10% stretch along their axis and 5% transverse (longitudinal) or 10% transverse and 5% along the cell axis (transverse). Myocytes can differentiate between these 2 stretching patterns, in this case with greater protein production when larger stretch is applied across the long axis of the cells.

applied via static stretch of the substrate. There is also evidence that cells can distinguish whether loading is altered during diastole or systole or both;[18] myocytes can elicit different signaling pathways depending on the timing of altered load during the cardiac cycle. It was also found that the orientation of stretch applied to aligned myocyte cultures affects the remodeling response:[19] static stretch along the myofiber axis had little effect on protein degradation, while stretch transverse to the cell axis altered myofibrillar organization and decreased protein degradation.

Recently, Gopalan and coworkers[20] have also examined the response of cultured myocytes to different patterns of stretch. Since cell shape and alignment, cell-matrix adhesion and cell-cell contact can all affect growth, and because mechanical strains in vivo are multiaxial and anisotropic, an in-vitro system was developed for engineering aligned, rod-shaped neonatal cardiac myocyte cultures. Microfabrication techniques were used to pattern collagen matrices in parallel lines on deformable silicone elastomers. Confluent, elongated aligned myocytes were produced which enabled the application of different strain patterns relative to a local cell axis (Fig. 1A). An elliptical cell stretcher applied 2:1 anisotropic strain statically to the elastic substrate, with the axis of greatest stretch (10%) either parallel or transverse to the myofibrils. After 24 hours, principal strain parallel to myocytes did not significantly alter myofibril accumulation or expression of atrial natriuretic factor (ANF), connexin-43 or N-cadherin compared with unstretched controls. In contrast, 10% strain transverse to the long axis of the cell resulted in upregulation of protein signal intensity by western blotting of ANF, connexin-43 and N-cadherin (Fig. 1B). Thus myocytes are able to differentiate the direction of external loads and show a differential hypertrophy and remodeling response depending on the type of loading. A similar result was found in skeletal muscle by Kumar et al,[21] where distinct signaling pathways are activated in response to mechanical stress applied axially and transverse to the muscle fibers. This becomes highly significant in the intact heart where the complex fiber architecture translates to different loading conditions on individual cells dependent on many factors including regional location, local extracellular force-transmitting environment, timing during the cardiac cycle, local material properties, among others.

Mechanotransduction

The molecular signaling processes involved in myocyte adaptation to altered loading conditions have been investigated in some detail,[22] but key aspects of the mechanical stimuli that evoke these responses remain unclear. Several hypotheses have been proposed for the mechanisms of mechanotransduction during the cardiac growth response.[23] The mechanical stimulus may be directly transmitted to the nucleus via cytoskeletal connections, or indirectly through a variety of signaling cascades.[10] Since it is likely that there is a direct mechanical coupling of external load to intracellular cytoskeletal structures via the cell membrane,[24,25] and the membrane deforms in response to external loads, it might be expected that the sarcolemma may play a role in the response to mechanical loads. Several potentially interacting candidates mediating mechanotransduction have been proposed including: activation of adenylate cyclase and cAMP-protein kinase;[26] increased activation of membrane bound phospholipase C leading eventually to increased protein kinase C activation;[27,28] mechanosensitive ion channels;[29] and increases in intracellular pH via activation of the Na^+/H^+ exchanger.[26] It has also been proposed that the extracellular matrix presents the stimulus to myocyte hypertrophy at focal adhesions on the cell membrane (the "costameres" of the cardiac myocyte) via an integrin-dependent or integrin-modulated pathway.[25,30]

Since integrins physically connect the cytoskeletal elements to the extracellular matrix,[31] they could be part of the mechanosensor that transmits mechanical signals to the cytoskeleton[32] and eventually to the nucleus. Integrins can modulate a variety of signaling cascades, and have recently been shown to act as mechanotransducers in cardiac fibroblasts,[33] and to participate in the hypertrophic response[34,35] and stress sensing[36] in myocytes. MacKenna et al[33] showed that extracellular signal-regulated kinase (ERK) and c-Jun NH2-terminal kinase (JNK) activation by stretch is matrix-specific in adult cardiac fibroblasts. By altering matrix composition and using function blocking peptides and/or antibodies to vary the receptors that wereh ligated while maintaining cell adhesion and spreading, ERK and JNK activation were shown to be integrin-dependent. In neonatal myocytes, the stretch response can also be partially blocked with antibodies against α1 and β1 integrin subunits (α1β1 is a collagen receptor), and by overlaying with laminin (which also ligates α1β1).[36] Overlaying with fibronectin also inhibited the rod-shaped phenotype suggesting that other integrins may be involved in myocyte shape regulation. Hence, integrin ligation can regulate cell phenotype and stretch responsiveness, and suggests a mechanism by which forces outside of the cell are transmitted to intracellular structures where mechanotransduction can occur.

Cardiac myocytes have another well-known length transducer: the sarcomere. The Frank-Starling mechanism in cardiac muscle is generally attributed to length-dependent activation, a process by which the sarcomere can detect changes in length and elicit the appropriate response, in this case increased contractile force.[37] Several mechanisms for this length dependent activation have been proposed, including lattice filament spacing,[38] modulation of the cooperativity of the calcium binding process to troponin C and cross-bridge formation,[39] and interactions of titin with actin and myosin.[40] Whichever of these mechanisms prove to be responsible for length-dependent activation, it is clear that sarcomeres can detect length changes, hence themselves act as a mechanotransducer. Besides the acute increase in contractile force, several mechanisms provide for increased intracellular calcium levels with stretch, for example, hormones released in response to mechanical stretch, such as angiotensin II, promote calcium release in the myocyte.[41] The elevated calcium levels can trigger molecular pathways leading to gene upregulation and cellular remodeling. Cardiomyocytes can detect intracellular calcium levels via calmodulin, which can trigger activation of the hypertrophy response.[3] Thus calcium-dependent signaling is one mechanism that myocytes can use to detect length changes and elicit a remodeling response.

Though numerous potential pathways have been proposed for the loading signal to eventually lead to gene regulation and protein expression, the actual mechanotransducer, converting the mechanical signal into a chemical one, is not known. If a structure such as the sarcomere is the sensor, it remains unclear whether the sarcomeric proteins would sense a length change (stretch or strain), or somehow transduce stress. Since a direct "stress sensor" is not known to exist,[42] it is likely that the cellular transducer is actually a strain or length transducer. This is certainly the case for man-made force gauges, which detect stress or force indirectly through the deformation of a compliant structure. Thus, the distinction between transducing stress and strain at the sensor level is probably related to the stiffness of the deformation-sensing element. In experiments by Kumar and colleagues in skeletal muscle, the same axial and transverse stress was applied to cells, but owing to the difference in fiber vs. transverse stiffness, the strain in the two directions was different and resulted in differential activation of signaling pathways.[21] This would implicate strain as the signal to which the myocytes respond to. Cells may even regulate their mechanotransduction pathways by acutely altering the local stiffness of protein complexes involved in these processes, for example by phosphorylation. It is apparent that the mechanosensor, and even the type of mechanical signal which is sensed, are still not well understood in the mechanotransduction pathway of the myocyte.

The Cytoskeleton and Its Role in Load Transduction

The structural coupling between the extracellular matrix and the cytoskeleton at focal adhesions (or costameres in vivo) suggests that external physical signals may be mechanically propagated via the cytoskeleton.[36] Although Sadoshima and coworkers[24] found that the myocyte cytoskeleton is not a likely candidate for mechanical transmission, since disruption of the microtubules and microfilaments did not prevent stretch-induced expression of immediate early genes, numerous studies have been conducted since, which do implicate the cytoskeleton in mechanotransduction. Wang et al proposed that mechanotransmission occurs across the cell surface via extracellular matrix-integrin-cytoskeletal connections.[25] Ingber and coworkers have suggested that acute reorganization of microtubules, intermediate filaments and microfilaments in response to external loads may be an initial step in cell remodeling.[43] These experiments were performed in endothelial cells, and it is not known if the same mechanism is active in cardiac myocytes. A mechanical link between the cytoskeleton and nucleus has also been shown,[44] which may provide a physical pathway for nuclear reception of external forces. Ezzell et al[45] have demonstrated that cells lacking vinculin, a Z disc associated protein, lost their ability to efficiently spread, and the mechanical stiffness of the integrin-cytoskeletal linkage was decreased. Other cytoskeletal proteins have also been implicated in the load-sensing pathway of myocyte hypertrophy.[46,47] For example, alterations in the giant protein titin correlate with reduced myocardial function in chronic heart failure.[48] In cardiomyopathy, an altered organization of titin and α-actinin has been found,[49] suggesting a possible role for the cytoskeleton in these pathologies. Olson et al[50] reported that a mutation in cytoskeletal actin is responsible for a familial human dilated cardiomyopathy. They speculated that defective transmission of force from contractile elements to the sarcolemma may be involved in the pathogenesis of heart failure due to this abnormality in the cytoskeleton. Disruption of other cytoskeletal proteins, such as desmin, plakoglobin, N-cadherin, plectin, and vinculin all produce a dilated cardiac phenotype with impaired function, either in fetal development or after birth.[46,47,51] Membrane associated cytoskeletal proteins such as those in the dystrophin complex have also been associated with cardiomyopathies.[52] These proteins connect intracellular structures with the extracellular matrix and thus are likely a force-transmitting component of the myocyte. Chien[53] has proposed that defects in the cytoskeletal component of the myocyte eventually result in a dilated cardiomyopathic phenotype. By contrast, mutations in the sarcomeric proteins generate a pattern of hypertrophic cardiomyopathy with preserved ventricular systolic function in addition to the myocyte hypertrophy and disarray.[54] Thus, there is evidence that cytoskeletal

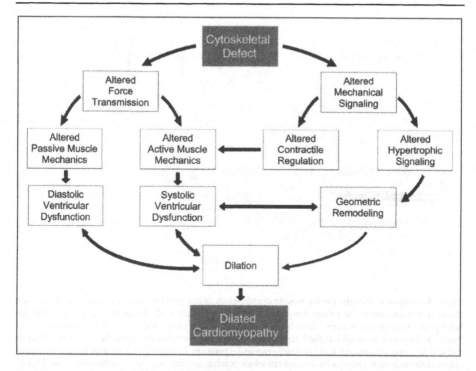

Figure 2. The two major possible pathways which could lead from a cytoskeletal defect to dilated cardiomy-opathy and heart failure. A defect in force transmission between the contractile proteins and the extracellular matrix could alter passive and active muscle mechanics as well as dysfunction on the ventricular level and lead to chamber dilation. Altered force transduction or mechanical signaling, on the other hand, could result in impaired calcium cycling and short-term contractile regulation or abnormal longer-term growth signaling, which in turn could lead to altered growth and remodeling of the ventricle and ultimately chamber dilation and failure.

proteins play a critical role in biomechanical signaling leading to compensatory hypertrophy and possibly ventricular dilation and heart failure.[6,55]

The various studies linking the cytoskeleton with dilated cardiomyopathy and heart failure have led to the hypothesis that there may exist a common pathway to cardiac dilation and failure, and the critical components of this "failure transition"[56] are likely to be structural them-selves or closely related to force transmitting elements. Several possible pathological pathways have been suggested through which dysfunction of a cytoskeletal protein could lead to heart failure. These pathways are illustrated in Figure 2. The lacking protein could play a structural role and mediate the force transmission between the contractile filaments and the extracellular matrix. A defect in force transmission could result in abnormal diastolic and systolic muscle mechanics, leading to ventricular dysfunction and dilation. Alternatively, the protein may have a signaling function and regulate force transduction or protein-protein interactions in the myocyte. In this case, the cytoskeletal defect may either result in abnormal contractile regula-tion, mediated by calcium handling in the cell, which in turn influences active muscle mechanics, or the defect may lead to altered hypertrophic and growth signaling, with the result of pathological geometric remodeling and ventricular dilation.

LIM Protein Deficiencies Lead to Cardiomyopathies

As outlined above, the myocyte cytoskeleton consists of numerous proteins, many of which have been associated with various forms of cardiac disease. Recent studies have indicated a

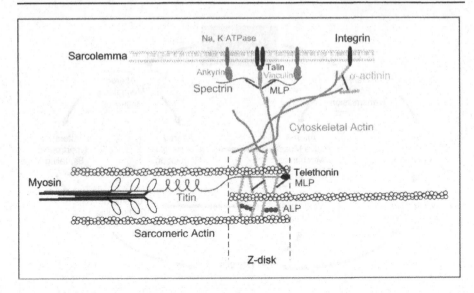

Figure 3. Components of the cardiomyocyte cytoskeleton. Select proteins with structural significance are shown in this schematic. For a more complete description, see Clark et al.[5] Forces are transmitted from the force-generating actin-myocyte cross-bridges through a complex pathway to the sarcolemma and trans-membrane proteins such as the integrins. From there forces are distributed to the extracellular matrix and to other myocytes to eventually produce ventricular pressure. Loads from the outside are also transmitted via these structural proteins back to the cell where mechanoperception and transduction occur. Muscle LIM protein (MLP) has been shown to bind to α-actinin, spectrin and telethonin, thus plays a critical role in force transmission in the cytoskeleton, particularly at the Z disc. The structural arrangement of the proteins could determine their relative importance in transverse vs. longitudinal mechanotransduction.

significant role of muscle-specific LIM domain proteins[57] in mechanical stress-sensing in the myocyte. LIM domain proteins make up a diverse group of regulatory and structural proteins.[58] Several of these LIM proteins are Z disc associated, and the Z disc of the cardiac myocyte is known to be an anchoring site for many structural proteins such as actin, α-actinin, titin and nebulin.[5] The LIM domain is known to be a protein binding interface,[59] thus it would be expected that proteins with a LIM domain would take on a structural or linker role at the Z disc and other locations in the cytoskeleton. Members of this LIM protein family expressed in muscle include muscle LIM protein (MLP),[60] enigma,[61] actinin-associated LIM protein (ALP),[62] cypher,[63] four and a half LIM-only protein FHL/SLIM,[64] and heart LIM protein (HLP).[65] Although the roles of most of these proteins in mechanotransduction are not known or may not exist, both ALP and MLP have been shown to play a role in cardiac remodeling and thus have a putative role in the mechanotransduction pathway.

The LIM domain protein ALP has been shown to localize in association with α-actinin-2 at the Z discs of myofibers (Fig. 3). α-actinin is a myofibrillar component of the cytoskeleton, one of various proteins that interconnect the thick and thin filaments of striated muscle. α-actinin links actin filaments laterally to each other and to the cell membrane at the Z disc.[66] ALP contains a PDZ domain, which binds to α-actinin-2 in muscle, and a single LIM domain. Interacting partners of the LIM domain have not been identified. PDZ domains are also involved with protein-protein interactions in the cytoskeleton[67] and may play a role in signal transduction.[68] Thus, ALP may play an important role in cytoskeletal structure, assembly and maintenance, and may be involved with signal transduction via the cytoskeleton. ALP has been shown to be predominantly localized in the right ventricle of the adult heart, although LacZ staining in the embryonic heart shows localization in the trabeculated areas of both LV and RV.[69] It was shown that ALP has an essential role in right ventricular morphogenesis. Lack of

ALP led to a decrease in trabeculation, chamber dilation and dysmorphogenesis of the embryonic RV. In the adult heart, the lack of ALP resulted in a mild form of right ventricular cardiomyopathy.[69] Although the need for ALP as a structural protein may be restricted to a certain time frame during embryonic development, there is evidence that this LIM domain protein is important during development when increased stresses are being transduced by the maturing cells and tissue.

Evidence of a genetic link between the cytoskeleton and dilated cardiomyopathy has been provided by studies of mice that harbor a deficiency in a muscle-specific LIM domain protein or MLP and display many features of human dilated cardiomyopathy.[70] Systolic performance of the tissue is decreased significantly in MLP deficient mice in terms of fractional and velocity of shortening of the LV wall and responsiveness to β-adrenergic stimulation. The minimum and maximum rates of change of LV pressure are reduced and LV end-diastolic pressure is increased, consistent with clinical LV pump failure. While several different factors have been attributed to the onset of the disease in man,[71] this mouse model does reproduce many of the clinical features of dilated cardiomyopathy, and eventually, many of the signs of heart failure. Very young MLP deficient mice did not show abnormalities in protein and myofibrillar components of the myocardium, suggesting that MLP plays a role during the physiologic growth processes[70] which are likely load dependent.

Both pathogenic pathways illustrated in Figure 2, via force transmission and via mechanical signaling, have been suggested to play a role in the development of dilated cardiomyopathy in MLP deficient mice.[72-75] Since MLP deficient mice display a disorganization of the sarcomeres with misaligned Z discs, which can be reversed by transfection of MLP, it was suggested that this LIM domain protein may serve as a scaffold and anchorage protein.[70] The finding that both LIM domains are required for MLP to function normally also supports the idea that MLP may be involved in force transmission or may serve as a component of a mechanical sensor to transduce hemodynamic forces into specific signaling responses. MLP is located at the Z disc of sarcomeres and is associated with α-actinin and telethonin (Fig. 3). Telethonin interacts with the end of titin facing the Z disc, therefore also called t-cap, and has itself been found to play a crucial role in pathogenesis in some patients with idiopathic dilated cardiomyopathy.[75] MLP's indirect association with titin supports the hypothesis that MLP may serve as a structural protein in force transmission, since titin has been identified as a major determinant of diastolic stress in cardiac muscle, especially at low stretch.[76] Telethonin has also been linked to a stretch-regulated potassium channel.[77] Perhaps through its interaction with telethonin, MLP may influence contractile signaling via the action potential and calcium handling. It has been suggested that the structural integrity of the cardiomyocytes was compromised in this mouse model preventing normal mechanical transduction of an overload stimulus and leading to a dilated cardiomyopathic state. Finally, MLP is also found near the costamere, a protein complex at the sarcolemma linking sarcomeres laterally at the Z disc between adjacent myofibers, where it colocalizes with spectrin.[72] Thus, it could be partly responsible for lateral force transmission across cell boundaries, particularly since occasional detachment of neighboring myocytes can be observed in MLP deficient myocardium.[78]

To investigate whether MLP plays a role in force transmission, isolated left ventricular papillary muscles from two-week-old MLP knockout and wildtype mice were mechanically tested.[75] Echocardiography in two-week-old knockout mice did not reveal signs of impaired cardiac performance. However, the stress-strain relationship in papillary muscles from MLP deficient mice displayed a decreased slope (Fig. 4). In other words, MLP knockout myocardium appeared substantially softer than controls. This could directly lead to an increased end-diastolic volume and dilation of the ventricle. The decreased stress could also alter the load-sensing regulation and lead to inadequate adaptation to the stress stimulus. A recent study investigated whether MLP may be directly involved in force transduction or the signaling cascade downstream.[75] Neonatal MLP deficient and control myocytes were isolated, plated on a membrane and stretched. Control myocytes increased the level of expression of the hypertrophy marker brain natriuretic peptide (BNP) in response to 10% stretch. Isolated myocytes from MLP

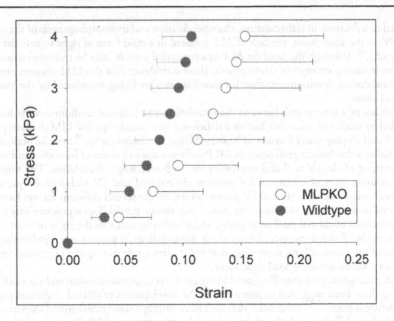

Figure 4. Diastolic mechanics in two-week-old MLP -/- hearts. Before the onset of chamber dilation and ventricular dysfunction, left ventricular papillary muscles from young MLP -/- hearts display a decreased diastolic stiffness compared to wildtype controls. This is shown by the right shift and decreased slope of the stress-strain relationship.

knockout mice were unable to respond to the stretch with an upregulation of BNP. However, upregulation of BNP could be induced hormonally using endothelin in MLP deficient myocytes. The investigators concluded that the role of MLP lies far upstream in the load signaling cascade, close to the actual stress-sensing mechanism. These functional changes in young MLP deficient myocardium before the onset of dilated cardiomyopathy are consistent with the decrease in myocardial stiffness being the initiating factor of the disease (due to an impairment of force transmission) and an inadequate compensation by concentric hypertrophy (due to the impairment in stress signaling).

The structural and mechanical signaling changes that result from lack of MLP in the young heart eventually lead to ventricular dilation and heart failure. Although the mechanisms of this process remain unclear, several of the specific functional consequences have been examined as outlined above. We have also found that passive mechanics and tissue structure are altered in the adult MLP -/- heart.[79] Passive mechanics of the myocardium are influenced by many factors[80] including geometry, material properties and boundary conditions. It is well known that the material properties of the myocardium depend on the local orientation of myocytes,[81] which have material anisotropy, and are a major determinant of passive stiffness. In the adult heart lacking MLP, there were significant changes in the local myocyte architecture, which are shown as changes in muscle fiber angle and laminar sheet orientation (Fig. 5). There were also changes in the passive mechanics that suggest a stiffer tissue. These changes include shifting of the passive pressurevolume relationship, as well as alterations in the local pressurestrain relations (Fig. 6). Based on these results, we concluded that disruption of the MLP component of the myocyte cytoskeleton leads to altered passive mechanical properties of the left ventricle during the development of ventricular dilation and systolic dysfunction. These functional differences exist in association with structural changes in the laminar sheet architecture in the left ventricle, which may be a mechanism behind the altered passive filling function. The passive dysfunction may be an important component of the overall loss of ventricular performance seen in this model of dilated cardiomyopathy.

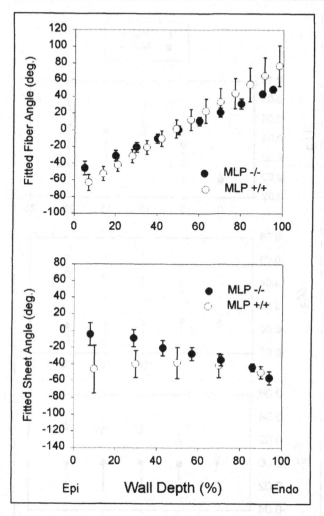

Figure 5. Fiber and sheet angles change in MLP deficient hearts. Lack of the MLP gene and (MLP -/-) leads to dilated cardiomyopathy and heart failure. These hearts show structural remodeling of transmural myocyte architecture compared to wild-type controls (+/+). The classic muscle fiber angle measure in the epicardial tangent plane shows modest changes (A), but the laminar sheet orientation shows substantial remodeling, particularly near the epicardium (B). (Reproduced with permission from Omens, et al.[79])

The changes seen in the adult myocardium of MLP knockout mice are a result of ventricular remodeling over the course of weeks or months. It is as yet unclear, whether the initial decrease and eventual increase in stiffness are directly related, and if so, whether the remodeling is due to intrinsically defective stress-sensing and stress-adaptation or due to a compensatory process in response to the decreased stiffness. In fact, elevated ventricular stiffness appears to be a common feature in both hypertrophy and cardiomyopathy.[82-85] Therefore, myocardial stiffening in this mouse model may conceivably occur as a part of the vicious circle of mechanical properties and cardiac performance leading to heart failure, and may be unrelated to the specific cause of the disease.

Mechanotransduction is typically considered to detect diastolic and systolic forces in the muscle fiber direction. As far as force generation in the myocyte is concerned, the axial

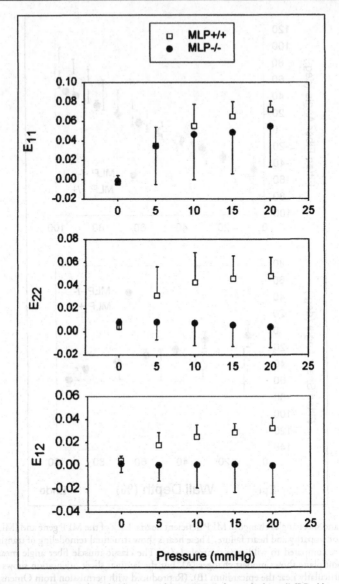

Figure 6. Passive mechanics are altered in hearts without MLP. In hearts lacking (MLP -/-), passive function has been altered. Pressure-strain relationships on the epicardium in the arrested hearts show significant changes in longitudinal (E_{22}) and in-plane shear (E_{12}) strain. Shear became essentially zero, indicating torsional dysfunction in the MLP -/- hearts. When rotated to fiber coordinates on the epicardium, the largest changes are seen in the cross-fiber component of passive strain. (Reproduced with permission from Omens et al.[79])

direction would be the most important since the sarcomere naturally generates forces parallel the orientations of the thin and thick filaments. However, forces in the direction transverse to the myofibrils are also thought to play a substantial role in normal ventricular contraction. It has been suggested that cross-fiber stress in the myocyte may be up to 40% of fiber stress.[86] These cross-fiber stresses are likely transmitted both from the sarcomere to the outside, and also from the outside to the load-sensing machinery inside the cell. Lateral forces are likely transmitted through the Z disc, not only because the Z disc is a major laterally oriented protein structure in the sarcomere, but also because sarcomeres are under normal conditions strictly aligned across myofibril boundaries and even cell boundaries. Since MLP is thought to be associated with the actin cytoskeleton both at the Z disk and near the sarcolemma (Fig. 3), this lateral linkage of sarcomeres suggests that MLP may be involved in cross-fiber force development and stress-sensing. Other structural proteins may also have direction dependent role in mechanotransduction related to their architectural arrangement. As discussed earlier, myocytes are able to distinguish between loading in different directions relative to the cell geometry and respond with different remodeling patterns, thus we suspect that mechanosensors in the cell may have a directional dependence. Further research is needed to test these possibilities.

Many different forms of heart failure with various etiologies have been described in patients, and the number of animal models of heart failure is also numerous.[87] We propose that the cytoskeletal disruption found in MLP and ALP deficient mice will result in dysfunctional stress regulation at the single myocyte level, which translates to ventricular dilation at the tissue level and eventual heart failure. Many of the structural proteins in the cytoskeleton have not been well characterized, and the signaling pathways that interact with these mechanosenors also need to be elucidated.

Acknowledgements

We kindly acknowledge the support of this work by National Heart, Lung, and Blood Institute Grants HL-43026, HL-64321and HL-46345 (K. Chien).

References

1. Braunwald E. Pathophysiology of Heart Failure. In: Braunwald E, ed. Heart Disease: A Textbook of Cardiovascular Medicine. 4th ed. Philadelphia: WB Saunders Co., 1992:393-418.
2. Sugden PH. Mechanotransduction in cardiomyocyte hypertrophy. Circulation 2001; 103:1375-1377.
3. Sussman MA, McCulloch AD, Borg TK. Dance band on the titanic: Biomechanical signaling in cardiac hypertrophy. Circ Res 2002; 91:888-898.
4. Ross RS. The extracellular connections: The role of integrins in myocardial remodeling. J Card Fail 2002; 8:S326-S331.
5. Clark KA, McElhinny AS, Beckerle MC et al. Striated muscle cytoarchitecture: An intricate web of form and function. Annu Rev Cell Dev Biol 2002; 18:637-706.
6. Hein S, Kostin S, Heling A et al. The role of the cytoskeleton in heart failure. Cardvasc Res 2000; 45:273-278.
7. Sadoshima J, Izumo S. Mechanotransduction in stretch-induced hypertrophy of cardiac myocytes. J Recept Res 1993; 13:777-794.
8. Grossman W. Cardiac hypertrophy: Useful adaptation or pathologic process? Am J Med 1980; 69:576-583.
9. Mann DL, Kent RL, Cooper IV G. Load regulation of the properties of adult feline cardiocytes: Growth induction by cellular deformation. Circ Res 1989; 64:1079-1090.
10. Reneman RS, Arts T, van Bilsen M et al. Mechanoperception and mechanotransduction in cardiac adaptation: Mechanical and molecular aspects. In: Sideman S, Beyar R, eds. Molecular and Subcellular Cardiology: Effects of Structure and Function. New York: Plenum Press, 1995:185-194.
11. Tavi P, Laine M, Weckström M et al. Cardiac mechanotransduction: From sensing to disease and treatment. Trend Pham Sci 2001; 55:254-260.
12. Anversa P, Ricci R, Olivetti G. Quantitative structural analysis of the myocardium during physiological growth and induced cardiac hypertrophy: A review. J Am Coll Cardiol 1986; 7:1140-1149.
13. Ross Jr J. Adaptations of the left ventricle to chronic volume overload. Circ Res 1974; 34:35:64-70.
14. Grossman W, Jones D, McLaurin LP. Wall stress and patterns of hypertrophy in the human left ventricle. J Clin Invest 1975; 56:56-64.
15. Smith SH, Bishop SP. Regional myocyte size in compensated right ventricular hypertrophy in the ferret. J Mol Cell Cardiol 1985; 17:1005-1011.
16. Sadoshima J, Jahn L, Takahashi T et al. Molecular characterization of the stretch-induced adaptation of cultured cardiac cells. J Biol Chem 1992; 267:10551-10560.
17. Simpson DG, Sharp WW, Borg TK et al. Mechanical regulation of cardiac myofibrillar structure. Ann NY Acad Sci 1995; 752:131-140.

18. Yamamoto K, Dang QN, Maeda Y et al. Regulation of cardiomyocyte mechanotransduction by the cardiac cycle. Circulation 2001; 103:1459-1464.
19. Simpson DG, Majeski M, Borg TK et al. Regulation of cardiac myocyte protein turnover and myofibrillar structure in vitro by specific directions of stretch. Circ Res 1999; 85:59-69.
20. Gopalan SM, Flaim C, Bhatia SN et al. Anisotropic stretch-induced hypertrophy in neonatal ventricular myocytes micropatterned on deformable elastomers. Bioeng Biotech 2003; 81:578-587.
21. Kumar A, Chaudhry I, Reid MB et al. Distinct signaling pathways are activated in response to mechanical stress applied axially and transversely to skeletal muscle fibers. J Biol Chem 2002; 277:46493-46503.
22. Yamazaki T, Yazaki Y. Molecular basis of cardiac hypertrophy. Z Kardiol 2000; 89:1-6.
23. Ruwhof C, van der Laarse A. Mechanical stress-induced cardiac hypertrophy: Mechanisms and signal transduction pathways. Cardiovasc Res 2000; 47:23-37.
24. Sadoshima J, Takahashi T, Jahn L et al. Roles of mechano-sensitive ion channels, cytoskeleton, and contractile activity in stretch-induced immediate-early gene expression and hypertrophy of cardiac myocytes. Proc Natl Acad Sci USA 1992; 89:9905-9909.
25. Wang N, Bulter JP, Ingber DE. Mechanotransduction across the cell surface and through the cytoskeleton. Science 1993; 260:1124-1127.
26. Watson P. Function follows form: Generation of intracellular signals by cell deformation. FASEB J 1991; 5:2013-2019.
27. Sadoshima J, Izumo S. Mechanical stretch rapidly activates multiple signal transduction pathways in cardiac myocytes: Potential involvement of an autocrine/paracrine mechanism. EMBO J 1993; 12:1681-1692.
28. Kumoro I, Katoh Y, Kaida T et al. Mechanical loading stimulates cell hypertrophy and specific gene expression in cultured rat cardiac myocytes. J Biol Chem 1991; 266:1265-1268.
29. Bustamante JO, Ruknudin A, Sachs F. Stretch-activated channels in heart cells: Relevance to cardiac hypertrophy. J Cardiovasc Pharmacol 1991; 17:S110-S113.
30. Juliano RL, Haskill S. Signal transduction from the extracellular matrix. J Cell Biol 1993; 120:577-585.
31. Schwartz MA, Schaller MD, Ginsberg MH. Integrins: Emerging paradigms of signal transduction. Ann Rev Cell Dev Biol 1995; 11:549-599.
32. Sadoshima J, Izumo S. The cellular and molecular response of cardiac myocytes to mechanical stress. Ann Rev Physiol 1997; 59:551-571.
33. MacKenna DA, Dolfi F, Vuori K et al. Extracellular signal-regulated kinase and c-Jun NH2-terminal kinase activation by mechanical stretch is integrin-dependent and matrix-specific in rat cardiac fibroblasts. J Clin Invest 1998; 101:301-310.
34. Keller RS, Shai SY, Babbitt CJ et al. Disruption of integrin function in the murine myocardium leads to perinatal lethality, fibrosis, and abnormal cardiac performance. Am J Pathol 2001; 158:1079-1090.
35. Ross RS, Pham C, Shai SY et al. Beta1 integrins participate in the hypertrophic response of rat ventricular myocytes. Circ Res 1998; 82:1160-1172.
36. Simpson DG, Terracio L, Terracio M et al. Modulation of cardiac myocyte phenotype in vitro by the composition and orientation of the extracellular matrix. J Cell Physiol 1994; 161:89-105.
37. Konhilas JP, Irving TC, De Tombe PP. Frank-Starling law of the heart and the cellular mechanisms of length-dependent activation. Pflugers Arch 2002; 445:305-310.
38. Fuchs F, Smith SH. Calcium, cross-bridges, and the frank-starling relationship. News Physiol Sci 2001; 16:5-10.
39. Dobesh DP, Konhilas JP, de Tombe PP. Cooperative activation in cardiac muscle: Impact of sarcomere length. Am J Physiol 2002; 282:H1055-H1062.
40. Cazorla O, Wu Y, Irving TC et al. Titin-based modulation of calcium sensitivity of active tension in mouse skinned cardiac myocytes. Circ Res 2001; 88:1028-1035.
41. Lipp P, Lavine M, Tovey SC et al. Functional InsP3 receptors that may modulate excitation-contraction coupling in the heart. Curr Biol 2000; 10:939-941.
42. Arts T, Prinzen FW, Snoeckx LHEH et al. A model approach to the adaptation of cardiac structure by mechanical feedback in the environment of the cell. In: Sideman S, Beyar R, eds. Molecular and Subcellular Cardiology: Effects of structure and function. New York: Plenum Press, 1995:217-228.
43. Ingber DE. Tensegrity: The architectural basis of cellular mechanotransduction. Ann Rev Physiol 1997; 59:575-599.
44. Maniotis AJ, Chen CS, Ingber DE. Demonstration of mechanical connections between integrins, cytoskeletal filaments, and nucleoplasm that stabilize nuclear structure. Proc Nat Acad Sci 1997; 94:849-854.

45. Ezzell RM, Goldmann WH, Wang N et al. Vinculin promotes cell spreading by mechanically coupling integrins to the cytoskeleton. Exp Cell Res 1997; 231:14-26.
46. Straub V, Campbell KP. Muscular distrophies and the dystrophin-glycoprotein complex. Curr Op Neuro 1997; 10:168-175.
47. Ruiz P, Birchmeier W. The plakoglobin knock-out mouse: A paradigm for the molecular analysis of cardiac cell junction formation. Trends Cardiovasc Med 1998; 8:97-101.
48. Hein S, Scholz D, Fujitani N et al. Altered expression of titin and contractile proteins in failing human myocardium. J Mol Cell Cardiol 1994; 26:1291-1306.
49. Kawaguchi N, Fujitani N, Schaper J et al. Pathological changes in myocardial cytoskeleton in cardiomyopathic hamster. Mol Cell Biochem 1995; 144:75-79.
50. Olson T, Michals V, Thibodean SN et al. Actin mutations in dilated cardiomyopathy: A heritable form of heart failure. Science 1998; 280:750-752.
51. Li D, Tapscoft T, Gonzalez O et al. Desmin mutation responsible for idiopathic dilated cardiomyopathy. Circulation 1999; 100:461-464.
52. Ortiz-Lopez R, Li H, Su J et al. Evidence for a dystrophin missense mutation as a cause of X-linked dilated cardiomyopathy. Circulation 1997; 95:2434-2440.
53. Chien KR. Stress pathways and heart failure. Cell 1999; 98:555-558.
54. Seidman CE, Seidman JG. Molecular genetics of inherited cardiomyopathies. In: Chien KR, ed. Molecular basis of heart disease. Philadelphia: WB Saunders Co., 1998:251-263.
55. Towbin JA. The role of cytoskeletal proteins in cardiomyopathies. Curr Opin Cell Biol 1998; 10:131-139.
56. Codd MB, Sugrue DD, Gersh BJ et al. Epidemiology of idiopathic dilated and hypertrophic cardiomyopathy: A population-based study in Olmsted County, Minnesota 1975-1984. Circulation 1989; 80:564-572.
57. Katz AM. Cytoskeletal abnormalities in the failing heart: Out on a LIM? Circulation 2000; 101:2672-2673.
58. Dawid IB, Toyama R, Taira M. LIM domain proteins. C R Acad Sci III 1995; 318:295-306.
59. Schmeichel KL, Beckerle MC. The LIM domain is a modular protein-binding interface. Cell 1994; 79:211-219.
60. Arber S, Halder G, Caroni P. Muscle LIM protein, a novel essential regulator of myogensis, promotes myogenic differentiation. Cell 1994; 79:221-231.
61. Guy PM, Kenny DA, Gil GN. The PDZ domain of the LIM protein enigma binds to beta-tropomyosin. Mol Biol Cell 1999; 10:1973-1984.
62. Xia H, Winokur ST, Luo W-L et al. Actinin-associated LIM protein: Identification of a domain interaction between PDZ and spectrin-like repeat motifs. J Cell Biol 1997; 139:507-515.
63. Zhou Q, Ruiz-Lozano P, Martone ME et al. Cypher, a striated muscle-restricted PDZ and LIM domain-containing protein, binds to alpha-actinin-2 and protein kinase C. J Biol Chem 1999; 274:19807-19813.
64. Chu PH, Ruiz-Lozano P, Zhou Q et al. Expression patterns of FHL/SLIM family members suggest important functional roles in skeletal muscle and cardiovascular system. Mech Dev 2000; 95:259-265.
65. Yu TS, Moctezuma-Anaya M, Kubo A et al. The heart LIM protein gene (Hlp), expressed in the developing and adult heart, defines a new tissue-specific LIM-only protein family. Mech Dev 2002; 116:187-192.
66. Pavalko FM, Otey CA. Role of adhesion molecule cytoplasmic domains in mediating interaction with cytoskeleton. Proc Soc Exper Biol Med 1994; 205:282-293.
67. Marfatia SM, Morais Cabral JH, Lin L et al. Modular organization of the PDZ domains in the human discs-large protein suggests a mechanism for coupling PDZ domain-binding proteins to ATP and the membrane cytoskeleton. J Cell Biol 1996; 135:753-766.
68. Brenman JE, Bredt DS. Synaptic signaling by nitric oxide. Curr Opin Neurobiol 1997; 7:374-378.
69. Pashmforoush M, Pomies P, Peterson KL et al. Adult mice deficient in actinin-associated LIM-domain protein reveal a developmental pathway for right ventricular cardiomyopathy. Nat Med 2001; 7:591-597.
70. Arber S, Hunter JJ, Ross Jr J et al. MLP-deficient mice exhibit a disruption of cardiac cytoarchitectural organization, dilated cardiomyopathy, and heart failure. Cell 1997; 88:393-403.
71. Bender JR. Idiopathic dilated cardiomyopathy: An immunologic, genetic, or infectious disease, or all of the above? Circulation 1991; 83:704-706.
72. Flick MJ, Konieczny SF. The muscle regulatory and structural protein MLP is a cytoskeletal binding partner of betaI-spectrin. J Cell Sci 2000; 113:1553-1564.
73. Ecarnot-Laubriet A, De Luca K, Vandroux D et al. Downregulation and nuclear relocation of MLP during the progression of right ventricular hypertrophy induced by chronic pressure overload. J Mol Cell Cardiol 2000; 32:2385-95.

74. Su Z, Yao A, Zubair I et al. Effects of deletion of muscle LIM protein on myocyte function. Am J Physiol 2001; 280:H2665-H2673.
75. Knöll R, Hoshijima M, Hoffman HM et al. The cardiac mechanical stretch sensor machinery involves a Z-disk complex that is defective in a subset of human dilated cardiomyopathy. Cell 2003; 111:943-956.
76. Wu Y, Cazorla O, Labeit D et al. Changes in titin and collagen underlie diastolic stiffness diversity of cardiac muscle. J Mol Cell Cardiol 2000; 32:2151-62.
77. Furukawa T, Ono Y, Tsuchiya H et al. Specific interaction of the potassium channel beta-subunit minK with the sarcomeric protein T-cap suggests a T-tubule-myofibril linking system. J Mol Biol 2001; 313:775-784.
78. Ehler E, Horowits R, Zuppinger C et al. Alterations at the intercalated disk associated with the absence of muscle LIM protein. J Cell Biol 2001; 153:763-772.
79. Omens JH, Usyk TP, Li Z et al. Muscle LIM protein deficiency leads to alterations in passive ventricular mechanics. Am J Physiol Heart Circ Physiol 2002; 282:H680-687.
80. McCulloch AD, Omens JH. Factors affecting the regional mechanics of the diastolic heart. In: Glass L, Hunter PJ, McCulloch AD, eds. Theory of Heart: Biomechanics, Biophysics and Nonlinear Dynamics of Cardiac Function. New York: Springer-Verlag, 1991:87-119.
81. Usyk TP, Mazhari R, McCulloch AD. Effect of laminar orthotropic myofiber architecture on regional stress and strain in the canine left ventricle. J Elas 2000; 61:143-164.
82. Thiedemann KU, Holubarsch C, Medugorac I et al. Connective tissue content and myocardial stiffness in pressure overload hypertrophy. Bas Res Cardiol 1983; 78:140-155.
83. Doering CW, Jalil JE, Janicki JS et al. Collagen network remodeling and diastolic stiffness of the rat left ventricle with pressure overload hypertrophy. Cardiovasc Res 1988; 22:686-695.
84. Conrad C, Brooks W, Hayes J et al. Myocardial fibrosis and stiffness with hypertrophy and heart failure in the spontaneously hypertensive rat. Circulation 1995; 91:161-170.
85. Emery JL, Omens JH. Mechanical regulation of myocardial growth during volume-overload hypertrophy in the rat. Am J Physiol 1998; 273:H1198-H1204.
86. Lin DH, Yin FC. A multiaxial constitutive law for mammalian left ventricular myocardium in steady-state barium contracture or tetanus. J Biomech Eng 1998; 120:504-517.
87. Iannini JP, Spinale FG. The identification of contributory mechanisms for the development and progression of congestive heart failure in animal models. J Heart Lung Transplant 1996; 15:1138-1150.

CHAPTER 6

Second Messenger Systems Involved in Heart Mechanotransduction

Hiroshi Hasegawa, Hiroyuki Takano, Yunzeng Zou, Hiroshi Akazawa and Issei Komuro*

Abstract

Mechanical stress can be considered one of the major stimuli that evoke hypertrophic responses including reprogramming of gene expression in cardiac myocytes. Therefore, it is important to understand how mechanical loading is sensed by cardiomyocytes and converted into intracellular biomechanical signals leading to cardiac hypertrophy. When mechanical stress is received it is converted also into biochemical signals inside the cells. The signal transduction pathway leading to an increase in protein synthesis is similar to the pathway which is known to be activated by various humoral factors such as growth factors, hormones and cytokines in many other cells. In this review we start with initiation of stress induced signaling and then concentrate on signalling vie MAP-kinase, JAK/STAT and ECM/integrin pathways. Although multiple cellular events which occur in cardiac myocytes in response to mechanical stretch have been clarified, many questions remain unanswered.

Introduction

Since mechanical stress such as pressure and volume overload is a major cause of cardiac hypertrophy, cardiac hypertrophy is formed in response to hemodynamic overload in order to normalize wall stress.[1] Cardiomyocytes are terminally differentiated and lose their ability to proliferate soon after birth whereas nonmyocytes such as fibroblasts, smooth muscle cells and endothelial cells can still proliferate even in the adult heart. Therefore, when the heart is exposed to various stresses such as hemodynamic overload and myocardial infarction, cardiomyocytes individually grow in cell size without an increase in cell number, while cardiac fibroblasts grow in cell number and produce extracellular matrix (ECM) proteins such as collagens and fibronectins.[2-5] Enlargement of cardiac myocytes and abnormal proliferation of cardiac fibroblasts with excessive accumulation of ECM proteins in myocardium, which are two of the features of left ventricular remodeling, lead ultimately to dysfunction of the heart. Cardiac hypertrophy may be observed in various cardiovascular diseases such as hypertension, valvular diseases, myocardial infarction and cardiomyopathy. Clinical studies have shown that prolonged cardiac hypertrophy is an independent risk factor for many cardiac diseases such as ischemic heart disease, arrhythmia and sudden death.[6] Mechanical stress has been considered to be one of the major stimuli that evoke hypertrophic responces including reprogramming of gene expression in cardiomyocytes.[7] However, it remains to be determined how mechanical

*Corresponding Author: Issei Komuro—Department of Cardiovascular Science and Medicine, Chiba University Graduate School of Medicine, 1-8-1 Inohana, Chuo-ku, Chiba city, Chiba 260-8670, Japan.

Cardiac Mechanotransduction, edited by Matti Weckström and Pasi Tavi.

loading is sensed by cardiomyocytes and is converted into intracellular biomechanical signals leading to cardiac hypertrophy. Therefore, it is extremely important to determine the molecular mechanisms of cardiac hypertrophy, which may help us to establish effective measures to prevent and treat it.

While a variety of intrinsic and extrinsic stimuli such as endocrine disorders, mechanical stress, ischemia, neurohumoral factors and cytokines can induce cardiac hypertrophy, mechanical stress is most important as an initial stimulus.[7,8] In the past decade, a great number of intracellular proteins have been characterized as important molecules to form multiple intracellular signal transduction pathways from hypertrophic stimuli to the formation of cardiac hypertrophy; these include G proteins, phospholipase C (PLC), phosphatidylinositol-3 (PI3) kinase, protein kinase C (PKC), protein kinase A (PKA), low-molecule GTPases (Ras, Rac, Rho), receptor and nonreceptor tyrosine kinases, mitogen-activated protein (MAP) kinase family including extracellular signal-regulated kinases (ERKs), p38MAP kinase and c-Jun NH2-terminal kinase (JNK), Ca^{2+}-binding proteins, gp130-signal transducer, the janus kinase (JAK)/signal transducer and activator of transcription (STAT) family, myosin light chain (MLC) kinase, integrins and probably others.[7,8] These molecules cooperate with each other to lead to the reprogramming of gene expression, increase in protein synthesis and induction of the cardiac hypertrophy. Calcium ion (Ca^{2+}), which regulates a number of cellular events such as contraction, fertilization, differentiation, growth and survival, is a key molecule for inducing hypertrophic responses of cardiomyocytes.[9] It is well known that stretch of cardiomyocytes enhances Ca^{2+} transients and modulates their contractility. Ca^{2+}/calmodulin (CaM) dependent protein kinases (CaMKs) and phosphatase calcineurin have attracted much attention as critical mediators of Ca^{2+} signaling for inducing cardiac hypertrophy. Both CaMKs and calcineurin[10-13] play important roles in the development of cardiac hypertrophy. Here we describe the various second messenger systems in cardiac hypertrophy, induced by mechanical stress.

Neurohumoral Factors Mediate Mechanical Stress-Induced Cardiac Hypertrophy

To investigate the signaling pathways that translate mechanical stress into cardiac hypertrophy, we developed an in vitro device by which mechanical stress can be imposed on cardiomyocytes cultured on deformable silicone dishes in a serum-free condition. Mechanical stretch of neonatal rat cardiomyocytes activates protein kinase cascade of phosphorylation such as PKC, Raf-1 kinase, MAP kinase and p90rsk, which are involved in re-expression of a number of genes including immediate early response genes (IEG) such as *c-myc* and *c-fos*[14,15] and fetal type genes such as *atrial natriuretic peptide (ANP)*, *skeletal α-actin (sk. α-actin)* and *β-myosin heavy chain (β-MHC)* followed by increased protein synthesis without DNA synthesis,[16-20] confirming that cardiac myocytes can respond to mechanical stress in the absence of neuronal and hormonal factors. When a mechanical stress is received by a cardiac myocyte, the signal may be spread in the cardiomyocyte by a protein kinase cascade in the order of PKC, Raf-1 kinase, MAP kinase kinase (MEK), MAP kinase and p90rsk to evoke the nuclear events.[20-22] Rho family, tyrosine kinases and gp130 also participate in the mechanical load-induced cardiac hypertrophy.[23] A short time course of activation of the protein kinase cascade actually contributes to cardiomyocyte hypertrophy. It may be of interest to explore how these signalings converge in the development of cardiac hypertrophy.

Although mechanical stretch itself is an initial stimulus for cardiac hypertrophy, accumulating evidence has suggested that mechanical stress induces cardiac hypertrophy by increasing the local production of humoral factors such as angiotensin II (AngII), endothelin-1 (ET-1) and catecholamine.[20-22] The protein kinase cascade is also activated by mechanical stress on cardiomyocytes and this event is very similar to that observed when growth factors and cytokines are added to many cell types. It was thus conceivable that cardiac myocytes and noncardiomyocytes such as fibroblasts, endothelial cells and smooth muscle cells secrete some hypertrophy-promoting factors following stretch, and that they induce cardiac hypertrophy by autocrine or paracrine mechanism.

The local renin-angiotensin system plays a critical role in the development of cardiac hypertrophy.[24,25] All components of the renin-angiotensin system such as angiotensinogen, renin and angiotensin converting enzyme (ACE) have been identified at both the mRNA and protein levels in the heart, and are increased in hypertrophied left ventricle of rats.[25] Subpressor doses of ACE inhibitors can prevent or cause regression of cardiac hypertrophy with no change in systemic systolic blood pressure.[26] An increase in left ventricular mass which was produced by abdominal aortic constriction, without significant increase in plasma renin activity, was completely prevented with an ACE inhibitor or AngII type 1 (AT1) receptor antagonists without any change in afterload.[26] These results strongly suggest that the local renin-angiotensin system but not the circulating one mediates pressure overload-induced cardiac hypertrophy.[27-29] AngII activates many protein kinases, such as ERKs,[30] JNK,[31,32] p38MAP kinase[33] and JAK/STAT,[34-36] and induces expression of IEG such as *c-fos* and *c-jun* genes as well as increases protein synthesis[30] in cardiac myocytes. Regarding the mechanism by which AngII induces activation of ERKs in cardiac myocytes, we and others have shown that AngII activates ERKs through AT1 receptor.[29,30] Activation of ERKs and up-regulation of *c-fos* expression by AngII are blocked by downregulation of PKC or a PKC inhibitor calphostin C, and AngII actually increases production of inositol phosphatases and activates PKC.[30] These signals elicited by AngII in cardiac myocytes are very similar to those evoked by mechanical stress,[37-40] so the involvement of AngII in the stretch-induced hypertrophic responces was examined using AT1-specific inhibitors. AngII also activated Raf-1 kinase and MEK, and the activation of these kinases is also suppressed by inhibition of PKC. Because PKC has been reported to directly activate Raf-1 kinase, AngII possibly activates ERKs through PKC, Raf-1 kinase and MEK.

ET-1 has been reported to be a strong inducer of cardiac hypertrophy.[41,42] Stretch-induced activation of ERKs are also suppressed by type A ET-1 (ETA) receptor antagonist,[43,44] suggesting that ET-1 is also involved in mechanical stress-induced hypertrophic responses. Since ET-1-induced activation of ERKs is also inhibited by down-regulation of PKC, ET-1 activates ERKs possibly through the pathway same as AngII.[43] Recently it has been reported that AngII does not have direct potent hypertrophy-promoting effects on cardiac myocytes, namely that AngII functions in cardiac hypertrophy by inducing ET-1 secretion from cardiac fibroblasts.[44] Cross-talk among different cell types and among many molecules might be important not only for maintaining its highly differentiated phenotype but also for inducing cardiac hypertrophy. The beneficial effects of ET-1 receptor antagonists on myocardial infarction highlighted the importance of ET-1 on cardiovascular remodeling.[45]

Recently, the importance of Gq protein in the development of pressure overload-induced cardiac hypertrophy was demonstrated using transgenic mice.[46] Class-specific inhibition of Gq-mediated signaling was produced in the heart of transgenic mice by targeted expression of a carboxyl-terminal peptide of the α subunit of Gq. When pressure overload was surgically produced, the transgenic mice developed significantly less ventricular hypertrophy than control mice. Although the prevention of cardiac hypertrophy was not complete in the transgenic mice, this study suggests that Gq-coupled receptors such as AT1, ETA receptor and other seven transmembrane receptors are important in the development of pressure overload-induced cardiac hypertrophy.

Cardiac hypertrophy is often associated with an increase in intracardiac sympathetic nerve activity and with elevated plasma catecholamines.[47] Treatment of cardiac myocytes with catecholamines not only changes their functions such as beating rates and contractile activity but also induces typical hypertrophic responses.[48-54] Prolonged infusion of subpressor doses of norepinephrine (NE) increases the mass of the myocardium and the thickness of the left ventricular wall, suggesting that NE has direct hypertrophic effects on cardiac myocytes without affecting afterload.[48] It has been reported that phenylephrine (PHE) evokes hypertrophic responses in cardiac myocytes of neonatal rats,[51] and that expression of constitutively active α-adrenergic receptor (AR) induces cardiac hypertrophy in adult mice.[50] Both in vivo and in vitro studies demonstrate that isoproterenol (ISO) also stimulates expression of proto-oncogenes in cardiomyocytes and induces cardiac hypertrophy.[49,52,53] We recently reported that the activation of protein kinase cascades followed by increased protein synthesis is induced by cAMP/PKA through β-AR as well as by PKC through α-AR.[54,55] NE-induced activation of ERKs is partly inhibited by either the α1-AR blocker prazosin or the β-AR blocker propranolol and is completely abolished by both blockers.[54] Activation of ERKs by PKA activators is dependent on trans-sarcolemmal Ca^{2+} influx and independent of PKC.[55] Moreover, PKA activators synergistically activate Raf-1 kinase and ERKs with PKC activators such as phorbol ester followed by increased protein synthesis.

Stimulation of β-AR usually activates cAMP/PKA through Gs protein. By examining the signal transduction pathway of ISO-induced activation of ERKs, we found that Gi protein (βγ subunits)/Src/Ras pathway is also required for ISO-induced ERKs activation.[56] How do two different pathways, cAMP/PKA pathway and Gi/Src/Ras pathway, converge at the ERK cascade? We further introduced β-AR mutant lacking phosphorylation sites for PKA into cardiac myocytes, and showed that inhibition of β-AR phosphorylation inhibits ISO-induced activation of ERKs. These results suggest that phosphorylation of β-AR by PKA and the change of coupling of the receptor from Gs to Gi play an important role in ISO-induced cardiomyocyte hypertrophy.[56]

Transactivation of the epidermal growth factor receptor (EGFR) has a role in GPCR-mediated signal transduction in various cells. Transactivation of EGFR is mediated by the EGFR ligand heparin-binding EGF (HB-EGF), which is cleaved from its membrane-anchored form (proHB-EGF) by a specific metalloproteinase. Shedding of HB-EGF resulting from metalloproteinase activation and subsequent transactivation of EGFR occurred when cardiomyocytes were stimulated by GPCR agonists, leading to cardiac hypertrophy. Both metalloproteinase inhibitor and HB-EGF neutralizing antibody abrogated the trans-activation of EGFR by PHE, AngII, or ET-1 in cultured rat neonatal cardiomyocytes. Metalloproteinase plays a crucial role in hypertrophic signaling of cardiomyocytes.[57]

The Mitogen-Activated Protein (MAP) Kinase Pathway

ERKs, the most extensively investigated members of the MAP kinase family, are the critical molecules for the development of cardiomyocyte hypertrophy.[58-60] Although activation of ERKs does not always lead to cardiomyocyte hypertrophy,[61] downregulation of ERKs by antisense oligodeoxynucleotides[59] or overexpression of ERK phosphatases[58,60] suppressed hypertrophic responses to growth stimuli in cardiac myocytes. ERKs have been shown to be involved in the expression of fetal type genes, such as *ANP* and *brain natriuretic peptide (BNP)* genes, and in protein synthesis.[62] Mechanical stress may activate ERKs and induce re-expression of IEG, such as *c-fos* and *c-jun*, and fetal type genes, such as *ANP*, *BNP* and *sk. α-actin* in cardiac myocytes.[62,63] Recently, ERKs were reported to activate phosphorylated heat- and acid-stable protein regulated by insulin, resulting in its dissociation from a cap-binding protein, eIF-4E, which initiates and increases cap-dependent protein synthesis.[64] Moreover, eIF-4E was demonstrated to be phosphorylated by electrically stimulated contraction of adult feline cardiomyocytes in vitro and acute pressure overload on canine hearts in vivo.[65] All these results suggest an important role of ERKs in mechanical stress-induced cardiac hypertrophy.

The JAK/STAT Pathway

JAKs are protein kinases associated with cytokine receptors that regulate signal transduction of these receptors[66] and STATs are latent transcription factors located in the cytoplasm which become activated by phosphorylation on a tyrosine residue. Binding of ligands, such as cardiotrophin-1 (CT-1), to their cytokine receptors leads to phosphorylation and activation of the receptor-JAK complex with subsequent recruitment of STATs and activation of STATs by phosphorylation.[67,68] The phosphorylated STATs dimerize, migrate into the nucleus, and bind response elements in the promoters of target genes to stimulate gene transcription. Activation of the JAK/STAT pathway by overload was mediated by the transmembrne glycoprotein gp130, and at least CT-1 and IL-6 were involved in activation of this pathway. gp130 may play a major role in signal transduction since it has been found that gp130 is an upstream activator of the JAK/STAT pathway as well as the ERK pathway.[69-71]

ECM-Integrin Pathway

Many studies have suggested that transduction of mechanical stress into biomechanical signals is largely mediated by a group of cell surface receptors called integrins, which link ECM to the cellular cytoskeleton, thus providing physical integration between the outside and the inside of the cell.[2,3,72] Therefore, transmembrane ECM receptors, such as the integrins, are also attractive candidates for mechanoreceptors. Integrins are heterodimeric transmembrane receptors and may couple ECM with the actin cytoskeleton.[73] A large extracellular domain of integrin receptor complexes binds various ECM proteins, while a short cytoplasmic domain has been shown to interact with the cytoskeleton in the cell.[72,73] Since cytoskeleton proteins can potentially regulate plasma membrane proteins such as enzymes and ion channels, it is speculated that mechanical stress could modulate these membrane-associated proteins and stimulate second messenger systems through the cytoskeleton. Therefore, integrins may be regarded as a mechanosensor that transmits mechanical signals to the cytoskeleton. Integrins can transmit signals not only through organizing the cytoskeleton but also by altering biochemical properties, such as the extent of tyrosine phosphorylation of a complex of proteins including focal adhesion tyrosine kinase (FAK).[72] FAK further activates various molecules such as Src, Fyn, Cas, Grb2 and PI3 kinase, and subsequently downstream proteins such as Ras, Rho, PKC and MAP kinases. A variety of integrins are expressed in the heart and participate in a number of biological processes including cardiac development and determination of morphology of cardiomyocytes.[73,74] When the outside-in signal of integrin was inhibited, stretch-induced activation of p38MAP kinase was strongly suppressed.[75] Activation of p38MAP kinase by stress was also abolished by overexpression of inhibitory molecules for FAK and Src. These results suggest that mechanical stress activates p38MAP kinase through the integrin-FAK-Src pathway in cardiac myocytes, providing evidence that integrins act as mechanoreceptors.[76-80] Recently, melusin, one of the candidate proteins to act as a biomechanical sensor in this passway, is reported. Melusin is a muscle-specific integrin β1-interacting protein, interacts with the cytosolic domain of integrin and is expressed exclusively in skeletal and cardiac muscle. Mice that are genetically deficient for the melusin gene do not develop cardiac hypertrophy induced by biomechanical stress through physical constriction of the aorta, but cardiac hypertrophy induced by neumohumoral factor through activation of Gq-coupled receptors was not influenced.[81] Moreover, mice that were melusin-deficient developed heart failure in response to biomechanical pressure overload. Although the downstream signaling pathway of melusin is unclear, this molecule may play the critical role in the development of cardiac hypertrophy induced by mechanical stress.

Ca²⁺ Regulates the Development of Cardiac Hypertrophy

It has been known that many cells rapidly respond to a variety of environmental stimuli by changes of ion channels in the plasma membrane.[82] Mechanosensitive ion channels have been observed with single-channel recordings in more than 30 cell types of prokaryotes, plants, fungi and all animals examined so far.[82] The activation of stretch-sensitive channels has been proposed as the transduction mechanism between mechanical stress and protein synthesis in cardiac hypertrophy.[83] Many kinds of stretch-sensitive ion channels, such as nonselective cation channels, K^+ channels, Cl^- channel, Ca^{2+} channels, Na^+/K^+ pump, Na^+/H^+ exchanger and Na^+/Ca^{2+} exchanger, have been found.[84-87] The stretch-sensitive channels allow the passage of the major monovalent physiological cations, Na^+ and K^+, and the divalent cation, Ca^{2+}. It is known that Ca^{2+} controls a number of cellular processes including cardiac hypertrophy. In response to stimuli, cardiomyocytes increase their cytosolic Ca^{2+} levels. With the use of a Ca^{2+}-binding fluorescent dye (fluo3) and the patch clamp technique, it was shown that mechanical stress-induced Ca^{2+} influx through stretch-sensitive channels leads to waves of Ca^{2+}-induced Ca^{2+} release.[88] Therefore, stretch-sensitive ion channels and exchangers are good candidates for the initial responder to mechanical stress. Hypertrophic stimulators including mechanical stress may increase intracellular Ca^{2+} levels and activate Ca^{2+}-dependent signaling systems in cardiac myocytes. Blocking of extra- and/or intracellular Ca^{2+} signaling abolished hypertrophic responses in cardiac myocytes.

The elevation of cytosolic Ca^{2+} levels is accomplished by influx from the external space and/ or release from internal stores such as the sarcoplasmic reticulum (SR).[89] In cardiac myocytes, Ca^{2+} influx through L-type Ca^{2+} channels induces a large Ca^{2+} release from SR through the ryanodine receptors (RyR) 2. Intracellular Ca^{2+} usually evokes cellular events by binding to Ca^{2+} binding proteins. A variety of Ca^{2+}-dependent signalings have been implicated in cardiac hypertrophy, but whether these signalings are independent or interdependent and whether there is specificity among them are unclear. CaM is a major Ca^{2+} binding protein present in all eukaryotic cells.[90] It has been reported that cardiomyocyte growth is specifically regulated by CaM concentrations and overexpression of CaM induces proliferative and hypertrophic growth of cardiomyocytes in transgenic mice.[91] Inhibition of CaM by specific inhibitor suppresses hypertrophic responses in cultured cardiomyocytes.[12] These findings provide important evidence for CaM to transduce Ca^{2+} signalings in cardiac myocytes. Ca^{2+}/CaM modulates many molecules through activation of functional molecules such as CaMKs and Ca^{2+}/CaM-dependent protein phosphatase calcineurin.[92,93] CaMKs are ubiquitous serine/threonine protein kinases involved in the regulation of diverse functions ranging from cell contraction, secretion and synaptic transmission to gene expression.[92] CaMKs have also been implicated in the development of cardiomyocyte hypertrophy. CaMKII, a predominant isoform of CaMKs expressed in the heart, regulates the expression of the *ANP* gene in contraction- and PHE-stimulated cardiac myocytes. We recently observed that CaMKII plays an important role in ET-1-induced activation of *β-MHC* promotor and increases in protein synthesis, myofibrillar organization and cell size in cardiomyocytes. On the other hand, Passier et al have indicated that activated CaMKI and CaMKIV also induce hypertrophy in cardiomyocytes in vitro and in vivo, and that the transcription factor myocyte enhancer factor-2 (MEF2) may be a downstream target for CaMKIV in the hypertrophic heart in vivo.[13] The same group further reported that CaMK stimulates MEF2 activity by relieving class II histone deacetylases (HDACs) from the DNA binding domain, and that the dissociation of HDACs by CaMK signaling may let MAP kinases to maximally stimulate MEF2 activity by phosphorylation of the transcription activation domain.[94] These findings demonstrate that synergistic activation of MEF2 involves convergence of CaMK and MAP kinase signaling pathways on different domains of the protein, providing a potential mechanism for transcriptional cross-talk between these signaling pathways in hypertrophic cardiomyocytes.

Calcineurin has been shown to play a pivotal role in neuronal functions and immune responses.[93] In particular, calcineurin has attracted great attention as a critical molecule that induces cardiac hypertrophy.[95] Overexpression of constitutively active calcineurin and of its downstream transcription factor- (NFAT3) induced marked cardiac hypertrophy in transgenic mice, and calcineurin inhibitors suppressed phenylephrine-, AngII- and ET-1-induced cardiomyocyte hypertrophy in vitro.[96] These observations suggest that an increase in intracellular Ca^{2+} levels could induce cardiac hypertrophy by activating calcineurin. However, the role of calcineurin in the pressure overload-induced cardiac hypertrophy has been unclear. To examine the function of calcineurin, we first used the in vitro stretch device. When cultured cardiac myocytes were stretched by 20% for 30 min, *BNP* gene was upregulated. The upregulation of *BNP* was suppressed by gadolinium (an inhibitor of stretch-sensitive ion channels) and attenuated by calcineurin inhibitors. In in vivo studies, many research groups, including ours, have reported that calcineurin plays a critical role in the development of pressure overload-induced cardiac hypertrophy, and several other groups reported an opposite conclusion.[97-100] Since many animals lost body weight and died possibly by severe side effects of calcineurin inhibitors in those studies, it might be difficult to reach a definite conclusion. To elucidate the precise role of calcineurin in the development of pressure overload-induced cardiac hypertrophy while avoiding drug toxicity, we have recently created transgenic mice which overexpress the dominant negative forms of calcineurin (D.N.CnA) only in the heart by using α-MHC promotor. In striking contrast to in vivo inhibition of calcineurin with CsA or FK506, the D.N.CnA transgene had no detrimental effects on nonoperated mice. After aortic banding, nontransgenic mice exhibited increased cardiac CnA activity, confirming the association between hemodynamic stress and CnA activation.[99] The increase in CnA activity was significantly reduced in the aorta-banded transgenic mice, demonstrating in vivo inhibition of myocardial CnA activity by the transgene. In a highly symmetrical experimental result, three weeks after banding there were greater decreases in several measures of cardiac hypertrophy, such as echocardiographic septal and posterior wall thicknesses, heart weight to body weight ratio, cardiomyocyte diameter, and the extent of left ventricular fibrosis. Interestingly, the reprogramming of expression of *c-fos*, *ANP*, *BNP* and *SERCA2* genes seen with pressure overload was attenuated in the D.N.CnA mice, but no change in *sk. α-actin* and *c-jun* mRNA were noted. Myocardial expression of D.N.CnA was thus effective in inhibiting cardiac hypertrophy at the whole-organ, cellular, and molecular levels after a pressure overload stimulus. Moreover, although it is suggested that calcineurin induces cardiac hypertrophy through its downstream transcription factor NFAT3 and interaction of NFAT3 with the cardiac zinc finger transcription factor GATA4, there may be other pathways through which calcineurin induces cardiac hypertrophy since we have observed that calcineurin was involved in the activation of ERKs in cardiac myocytes. A similar transgenic approach, with different CnA-inhibitory peptides, was recently reported.[100] Two transgenic mouse strains were created, which expressed the ΔCain and ΔAKAP79 peptides previously shown to inhibit AngII- and PHE-induced cardiomyocyte hypertrophy. Single- and double-transgene-copy ΔCain- and ΔAKAP79-transgenic mice were apparently normal at baseline. When exposed to pressure overload, ΔCain or ΔAKAP79 transgenic mice showed reduced cardiac CnA activity and hypertrophy. An especially strong feature of the report is the use of in vivo adenoviral infection of rat myocardium to assess the effects of ΔCain, independent of developmental perturbations that are inevitable with transgenic expression regulated by the α-MHC promoter. In effect, rats were treated with adenoviral ΔCain "gene therapy" that targeted CnA and then underwent aortic banding. Pressure overload-induced the normal increase in CnA activity was abolished by adenoviral ΔCain, and hypertrophy was diminished. These results show the potential for inhibition of CnA to modify reactive myocardial hypertrophy. Similarly, Rothermel et al[101] described the effects of transgenic expression of a truncated form of the endogenous CnA-inhibitory protein, myocyte-enriched CnA-inhibitory protein 1 (hMCIP1). This natural inhibitor for CnA is highly expressed in striated muscle and is transcriptionally upregulated as a consequence of CnA activation. Thus, hMCIP1 represents an endogenous negative regulator for myocyte CnA activity. Cardiac-specific expression of hMCIP1 inhibited cardiac hypertrophy, fetal gene expression, and progression to dilated cardiomyopathy that otherwise result from expression of a constitutively active form of CnA. Expression of the hMCIP1 transgene also inhibited hypertrophic responses to β-AR stimulation or exercise training. These results reveal that an endogenous, naturally regulated inhibitor of myocardial CnA is effective in modulating cardiac hypertrophy resulting from unrestrained CnA activity, catecholamine excess, and exercise. Calcineurin is critically involved in load-induced cardiac hypertrophy.[103-109]

Conclusions

Once mechanical stress is received and converted into biochemical signals, the signal transduction pathway leading to an increase in protein synthesis which is similar to the pathway which is known to be activated by various humoral factors such as growth factors, hormones and cytokines in many other cells (Fig. 1). Recently, yeast genes encoding members of MAP kinase have been isolated by complementation of yeast mutations as an essential protein for restoring the osmotic gradient across the cell membrane in response to increased external osmolarity.[110] This result suggests that mechanical stress-induced intracellular signal transduction pathways are highly conserved in evolution.

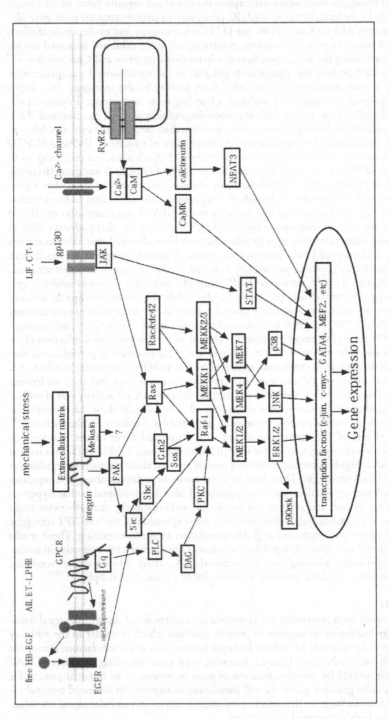

Figure 1. Mechanical stress-induced activation of intracellular signal transduction pathways leading to formation of the cardiac hypertrophy. Mechanical stress acting on cardiac myocytes stimulates mechanoreceptors such as stretch-activated ion channels (SA channels) including Ca^{2+} channels and exchangers, integrins, receptor tyrosine kinases and probably others, induces release of humoral factors such as AngII, ET-1, PHE and cytokines which act on cardiomyocytes by the autocrine and/ or paracrine mechanism, elevates intracellular Ca^{2+} levels through extracellular Ca^{2+} influx and/or intracellular Ca^{2+} release, activates multiple intracellular signal transduction pathways, and then results in increases in protein synthesis and reprogramming of specific gene expression, leading to formation of cardiac hypertrophy.

Although multiple cellular events which occur in cardiac myocytes in response to mechanical stretch have been clarified, many questions remain unanswered. First, how is mechanical stress converted into biochemical signals? In other words, what is the mechanoreceptor or the transducer for mechanical stress in cardiac myocytes? Although some molecules such as ion channels, exchangers and integrins seem to be candidates for the mechanoreceptor as mentioned above, it still needs to be defined which is the most important. Second, mechanical stress may stimulate secretion of some growth factors and cytokines, which generate multiple intracellular signals as a secondary event. The mechanism causing these secretions also needs to be identified. Third, a great number of intracellular signalings have been clearly shown to form transduction pathways leading to the formation of cardiac hypertrophy. Do these signalings function independently or synergistically? Is there a switch to control the processes? Further studies are necessary to identify them. Fourth, calcineurin is involved in not only the induction of cardiac hypertrophy but also the protection of cardiomyocytes from apoptotic death.[111] In contrast, there are many reports indicating that calcineurin induces cell apoptosis.[112] We also observed recently that ISO induces apoptosis through calcineurin activation in cardiac myocytes.[113] Moreover, a report indicated that calcineurin activation could either induce or suppress apoptosis.[114] Since cardiomyocyte apoptosis is a cause of heart failure,[115] it becomes important to determine the specific mechanism by which calcineurin induces or suppresses apoptosis of cardiomyocytes. Calcineurin induces cardiac hypertrophy through transcription factor NFAT3 and interaction of NFAT3 with GATA4,[108] however, it is still uncertain whether NFAT3 is essential for the calcineurin-dependent trophic pathway. Moreover, the calcineurin pathway may interconnect with other pathways such as PKC, JNK and ERKs.[111] How these events occur thus remains unclear. The answers to these questions should pave the way to understand the definite mechanism of mechanical stress-induced cardiac hypertrophy and to ultimately clarify the mechanism by which adaptive cardiac hypertrophy deteriorates into congestive heart failure.

References

1. Frohlich ED. Cardiac hypertrophy in hypertension. N Engl J Med 1987; 317:831-833.
2. Weber KT, Janicki JS, Shroff SG et al. Collagen remodeling of the pressure overloaded hypertrophied nonhuman primate myocardium. Circ Res 1988; 62:757-765.
3. Bashey RI, Donnelly M, Insinga F et al. Growth properties and biochemical characterization of collagens synthesized by adult rat heart fibroblasts in culture. J Mol Cell Cardiol 1992; 24:691-700.
4. Brilla CG, Maisch B. Regulation of the structural remodelling of the myocardium: From hypertrophy to heart failure. Eur Heart J 1994; 15:45-52.
5. Weber KT, Brilla CG. Pathological hypertrophy and cardiac interstitium. Fibrosis and renin-angiotensin-aldosterone system. Circulation 1991; 83:1849-1865.
6. Levy D, Garrison RJ, Savage DD et al. Prognostic implications of echocardiographically determined left ventricular mass in the Framingham heart study. N Engl J Med 1990; 322:1561-1566.
7. Komuro I, Yazaki Y. Control of cardiac gene expression by mechanical stress. Annu Rev Physiol 1993; 55:55-75.
8. Chien KR, Grace AA, Hunter JJ. Molecular biology of cardiac hypertrophy and heart failure. In: Chien KR, ed. Molecular Basis of Cardiovascular Disease. Philadelphia: Saunders Co., 1998:211-250.
9. Berridge MJ, Bootman MD, Lipp P. Calcium - a life and death signal. Nature 1998; 395:645-648.
10. McDonough PM, Glembotski CC. Induction of atrial natriuretic factor and myosin light chain-2 gene expression in cultured ventricular myocytes by electrical stimulation of contraction. J Biol Chem 1992; 267:11665-11668.
11. Ramirez MT, Zhao XL, Schulman H et al. The nuclear δB isoform of Ca^{2+}/calmodulin-dependent protein kinase II regulates atrial natriuretic factor gene expression in ventricular myocytes. J Biol Chem 1997; 272:31203-31208.
12. Zhu W, Zou Y, Shiojima I et al. Ca^{2+}/calmodulin-dependent kinase II and calcineurin play critical roles in endothelin-1-induced cardiomyocyte hypertrophy. J Biol Chem 2000; 275:15239-15245.
13. Passier R, Zeng H, Frey N et al. CaM kinase signaling induces cardiac hypertrophy and activates the MEF2 transcription factor in vivo. J Clin Invest 2000; 105:1395-1406.
14. Komuro I, Kurabayashi M, Takaku F et al. Expression of cellular oncogenes in the myocardium during the developmental stage and pressure overloaded hypertrophy of the rat heart. Circ Res 1988; 62:1075-9.

15. Komuro I, Kaida T, Shibazaki Y et al. Stretching cardiac myocytes stimulates protooncogene expression. J Biol Chem 1990; 265:3595-8.
16. Siri FM, McNamara JJ. Effects of sympathectomy on heart size and function in aortic-constricted rats. Am J Physiol 1987; 252:H442-H447.
17. Peterson MB, Lesch M. Protein synthesis and amino acid transport in isolated rabbit right ventricular muscle. Circ Res 1972; 31:317-327.
18. Cooper G, Kent RL, Uboh CE et al. Hemodynamic versus adrenergic control of cat right ventricular hypertrophy. J Clin Invest 1985; 75:1403-1414.
19. Yamazaki T, Tobe K, Hoh E et al. Mechanical loading activates mitogen-activated protein kinase and S6 peptide kinase in cultured rat cardiac myocytes. J Biol Chem 1993; 268:12069-76.
20. Yamazaki T, Komuro I, Yazaki Y. Molecular mechanism of cardiac cellular hypertrophy by mechanical stress. J Mol Cell Cardiol 1995; 27:133-40.
21. Yamazaki T, Komuro I, Yazaki Y. Signalling pathways for cardiac hypertrophy. Cell Signal 1998; 10:693-698.
22. Sugden PH, Clerk A. Cellular mechanisms of cardiac hypertrophy. J Mol Med 1998; 76:725-746.
23. Aikawa R, Komuro I, Yamazaki T et al. Rho family small G proteins play critical roles in mechanical stress-induced hypertrophic responses in cardiac myocytes. Circ Res 1999; 84:458-466.
24. Sadoshima J, Izumo S. The cellular and molecular response of cardiac myocytes to mechanical stress. Annu Rev Physiol 1997; 59:551-571.
25. Baker KM, Booz GW, Dostal DE. Cardiac actions of angiotensin II: Role of an intracardiac renin-angiotensin system. Annu Rev Physiol 1992; 54:227-241.
26. Linz W, Schoelkens BA, Ganten D. Converting enzyme inhibitor specifically prevents development and induces the regression of cardiac hypertrophy in rats. Clin Exp Hypertens 1989; 11:1325-1350.
27. Kojima M, Shiojima I, Yamazaki T et al. Angiotensin II receptor antagonist TCV-116 induces regression of hypertensive left ventricular hypertrophy in vivo and inhibits intracellular signaling pathway of stretch-mediated cardiomyocyte hypertrophy in vitro. Circulation 1994; 89:2204-2211.
28. Bruckschlegel G, Holmer SR, Jandeleit K et al. Blockade of the renin-angiotensin system in cardiac pressure overload hypertrophy in rats. Hypertension 1995; 25:250-259.
29. Yamazaki T, Komuro I, Kudoh S et al. Angiotensin II partly mediates mechanical stress-induced cardiac hypertrophy. Circ Res 1995; 77:258-265.
30. Sadoshima J, Xu Y, Slayter HS et al. Autocrine release of angiotensin II mediates stretch-induced hypertrophy of cardiac myocytes in vitro. Cell 1993; 75:977-984.
31. Kudoh S, Komuro I, Mizuno T et al. Angiotensin II stimulates c-Jun NH2-terminal kinase in cultured cardiac myocytes of neonatal rats. Circ Res 1997; 80:139-146.
32. Izumi Y, Kim S, Zhan Y et al. Important role of angiotensin II-mediated c-Jun NH(2)-terminal kinase activation in cardiac hypertrophy in hypertensive rats. Hypertension 2000; 36:511-516.
33. Pellieux C, Sauthier T, Aubert JF et al. Angiotensin II-induced cardiac hypertrophy is associated with different mitogen-activated protein kinase activation in normotensive and hypertensive mice. J Hypertens 2000; 18:1307-1137.
34. Pan J, Fukuda K, Kodama H et al. Role of angiotensin II in activation of the JAK/STAT pathway induced by acute pressure overload in the rat heart. Circ Res 1997; 81:611-617.
35. Hunt RA, Bhat GJ, Baker KM. Angiotensin II-stimulated induction of sis-inducing factor is mediated by pertussis toxin-insensitive Gq proteins in cardiac myocytes. Hypertension 1999; 34:603-608.
36. McWhinney CD, Dostal D, Baker K. Angiotensin II activates Stat5 through Jak2 kinase in cardiac myocytes. J Mol Cell Cardiol 1998;30:751-761.
37. Zou Y, Komuro I, Yamazaki T et al. Cell type-specific angiotensin II-evoked signal transduction pathways. Critical roles of Gβγ subunit, Src family, and Ras in cardiac fibroblasts. Circ Res 1998; 82:337-345.
38. Zou Y, Komuro I, Yamazaki T et al. Protein kinase C, but not tyrosine kinases or Ras, plays a criticle role in angiotensin II-induced activation of Raf-1 kinase and extracellular signal-regulated protein kinases in cardiac myocytes. J Biol Chem 1996; 271:33592-33597.
39. Sadoshima J, Izumo S. The heterotrimetric Gq protein-coupled angiotensin II receptor activates p21ras via the tyrosine kinase-Shc-Grb2-Sos pathway in cardiac myocytes. EMBO J 1996; 15:775-787.
40. Kudoh S, Komuro I, Hiroi Y et al. Mechanical stretch induces hypertrophic responses in cardiac myocytes of angiotensin II type 1a receptor knockout mice. J Biol Chem 1998; 273:24037-24043.
41. Ito H, Hirata Y, Hiroe M et al. Endothelin-1 induces hypertrophy with enhanced expression of muscle-specific genes in cultured neonatal rat cardiomyocytes. Circ Res 1991; 69:209-215.
42. Yamazaki T, Komuro I, Kudoh S et al. Endothelin-1 is involved in mechanical stress-induced cardiomyocyte hypertrophy. J Biol Chem 1996; 271:3221-3228.

43. Yamazaki T, Komuro I, Zou Y et al. Hypertrophic responses of cardiomyocytes induced by endothelin-1 through the protein kinase C-dependent but Src and Ras-independent pathways. Hypertens Res 1999; 22:113-119.
44. Harada M, Itoh H, Nakagawa O et al. Significance of ventricular myocytes and nonmyocytes interaction during cardiocyte hypertrophy: Evidence for endothelin-1 as a paracrine hypertrophic factor from cardiac nonmyocytes. Circulation 1997; 96:3737-3744.
45. Sakai S, Miyauchi T, Kobayashi M et al. Inhibition of myocardial endothelin pathway improves long-term survival in heart failure. Nature 1996; 384:353-355.
46. Akhter SA, Luttrell LM, Rockman HA et al. Targeting the receptor-Gq interface to inhibit in vivo pressure overload myocardial hypertrophy. Science 1998; 280:574-577.
47. Zou Y, Yao A, Zhu W et al. Isoproterenol activates extracellular signal -regulated protein kinases in cardiomyocytes through calcineurin. Circulation 2001; 104:102-108.
48. Laks MM, Morady F, Swan HJC. Myocardial hypertrophy produced by chronic infusion of subhypertensive doses of norepinephrine in the dogs. Chest 1973; 64:75-78.
49. Brand T, Sharma HS, Schaper W. Expression of nuclear proto-oncogenes in isoproterenol-induced cardiac hypertrophy. J Mol Cell Cardiol 1993; 25:1325-1337.
50. Milano CA, Dolber PC, Rockman HA et al. Myocardial expression of a constitutively active α1B-adrenergic receptor in transgenic mice induces cardiac hypertrophy. Proc Natl Acad Sci USA 1994; 91:10109-10113.
51. Thorburn A. Ras activity is required for phenylephrine-induced activation of MAP kinase in cardiac muscle cells. Biochem Biophys Res Commun 1994; 205:1417-1422.
52. Slotkin TA, Lappi SE, Seidler FJ. β-adrenergic control of c-fos expression in fetal and neonatal rat tissues: Relationship to cell differentiation and teratogenesis. Tox Appl Pharm 1995; 133:188-195.
53. Bogoyevitch MA, Andersson MB, Gillespie BJ et al. Adrenergic receptor stimulation of the mitogen-activated protein kinase cascade and cardiac hypertrophy. Biochem J 1996; 314:115-121.
54. Yamazaki T, Komuro I, Zou Y et al. Norepinephrine induces the raf-1 kinase/MAP kinase cascade through both α1- and β-adrenoceptors. Circulation 1997; 95:1260-1268.
55. Yamazaki T, Komuro I, Zou Y et al. Protein kinase A and protein kinase C synergistically activate the Raf-1 kinase/mitogen-activated protein kinase cascade in neonatal rat cardiomyocytes. J Mol Cell Cardiol 1997; 29:2491-2501.
56. Zou Y, Komuro I, Yamazaki T et al. Both Gs and Gi proteins are critically involved in isoproterenol-induced cardiomyocyte hypertrophy. J Biol Chem 1999; 274:9760-9770.
57. Asakura M, Kitakaze M, Takashima S et al. Cardiac hypertrophy is inhibited by antagonism of ADAM12 processing of HB-EGF: Metalloproteinase inhibitors as a new therapy. Nat Med 2002; 8:35-40.
58. Thorburn J, Carlson M, Mansour SJ et al. Inhibition of a signaling pathway in cardiac muscle cells by active mitogen-activated protein kinase kinase. Mol Biol Cell 1995; 6:1479-1490.
59. Glennon PE, Kaddoura S, Sale EM et al. Depletion of mitogen-activated protein kinase using an antisense oligodeoxynucleotide approach downregulates the phenylephrine-induced hypertrophic response in rat cardiac myocytes. Circ Res 1996; 78:954-961.
60. Fuller SJ, Davies EL, Gillespie BJ et al. Mitogen-activated protein kinase phosphatase 1 inhibits the stimulation of gene expression by hypertrophic agonists in cardiac myocytes. Biochem J 1997; 15:313-319.
61. Post GR, Goldstein D, Thuerauf DJ et al. Dissociation of p44 and p42 mitogen-activated protein kinase activation from receptor-induced hypertrophy in neonatal rat ventricular myocytes. J Biol Chem 1996; 271:8452-8457.
62. Page C, Doubell AF. Mitogen-activated protein kinase (MAPK) in cardiac tissues. Mol Cell Biochem 1996; 157:49-57.
63. Harada K, Komuro I, Zou Y et al. Acute pressure overload could induce hypertrophic responses in the heart of angiotensin II type 1a knockout mice. Circ Res 1998; 82:779-85.
64. Pause A, Belsham GJ, Gingras AC et al. Insulin-dependent stimulation of protein synthesis by phosphorylation of a regulator of 5'-cap function. Nature 1994; 371:762-767.
65. Wada H, Ivester CT, Carabello BA et al. Translational initiation factor eIF-4E. J Biol Chem 1996; 271:8359-8364.
66. Ihle JN. Cytokine receptor signalling. Nature 1995; 377:591-594.
67. Mascareno E, Dhar M, Siddiqui MA. Signal transduction and activator of transcription (STAT) protein-dependent activation of angiotensinogen promoter: A cellular signal for hypertrophy in cardiac muscle. Proc Natl Acad Sci USA 1998; 95:5590-5594.
68. Pan J, Fukuda K, Saito M et al. Mechanical stretch activates the JAK/STAT pathway in rat cardiomyocytes. Circ Res 1999; 84:1127-1136.

69. Uozumi H, Hiroi Y, Zou Y et al. gp130 plays a critical role in pressure overload-induced cardiac hypertrophy. J Biol Chem 2001; 276:23115-23119.
70. Hirota H, Chen J, Betz UA et al. Loss of a gp130 cardiac muscle cell survival pathway is a critical event in the onset of heart failure during biomechanical stress. Cell 1999; 97:189-98.
71. Kunisada K, Hirota H, Fujio Y et al. Activation of JAK-STAT and MAP kinases by leukemia inhibitory factor through gp130 in cardiac myocytes. Circulation 1996; 94:2626-2632.
72. Juliano RL, Haskill S. Signal transduction from the extracellular matrix. J Cell Biol 1993; 120:577-85.
73. Hynes RO. Integrins: Versatility, modulation, and signaling in cell adhesion. Cell 1992; 69:11-25.
74. Hamasaki K, Mimura T, Furuya H et al. Stretching mesangial cells stimulates tyrosine phosphorylation of focal adhesion kinase pp125FAK. Biochem Biophys Res Commun 1995; 212:544-549.
75. Aikawa R, Nagai T, Kudoh S et al. Integrins play a critical role in mechanical stress-induced p38 MAPK activation. Hypertension 2002; 39:233-8.
76. Ingber D. Integrins as mechanochemical transducer. Curr OpinCell Biol 1991; 3:841-848.
77. Simpson DG, Terracio L, Terracio M et al. Modulation of cardiac myocyte phenotype in vitro by the composition and orientation of the extracellular matrix. J Cell Physiol 1994; 161:89-105.
78. Ross RS. The extracellular connections: The role of integrins in myocardial remodeling. J Card Fail 2002; 8:S326-31.
79. Parsons JT. Integrin-mediated signalling: Regulation by protein tyrosine kinases and small GTP-binding proteins. Curr Opin Cell Biol 1996; 8:146-152.
80. Buck CA, Baldwin HS, DeLisser H et al. Cell adhesion receptors and early mammalian heart development: An overview. Sciences de la Vie 1993; 316:838-859.
81. Brancaccio M, Fratta L, Notte A et al. Melusin, a muscle-specific integrin beta1-interacting protein, is required to prevent cardiac failure in response to chronic pressure overload. Nat Med 2003; 9:68-75.
82. Morris CE. Mechanosensitive ion channels. J Membrane Biol 1990; 113:93-107.
83. Bustamante JO, Ruknudin A, Sachs F. Stretch-activated channels in heart cells: Relevance to cardiac hypertrophy. J Cardiovasc Pharm 1991; 17:S110-S113.
84. Craelius W, Chen V, el-Sherif N. Stretch activated ion channels in ventricular myocytes. Bioscience Rep 1988; 8:407-414.
85. Lederer WJ, He S, Luo S et al. The molecular biology of the Na^+-Ca^{2+} exchanger and its functional roles in heart, smooth muscle cells, neurons, glia, lymphocytes, and nonexcitable cells. Ann NY Acad Sci 1996; 779:7-17.
86. Yamazaki T, Komuro I, Kudoh S et al. Role of ion channels and exchangers in mechanical stretch-induced cardiomyocyte hypertrophy. Circ Res 1998; 82:430-437.
87. Sigurdson W, Ruknudin A, Sachs F. Ca^{2+} imaging of mechanically induced fluxes in tissue-cultured chick heart: Role of stretch-activated ion channels. Am J Physiol 1992; 262:H1110-H1115.
88. Nabauer M, Morad M. Ca^{2+}-induced Ca^{2+} release as examined by photolysis of caged Ca^{2+} in single ventricular myocytes. Am J Physiol 1990; 258:C189-C193.
89. Sham JS, Cleemann L, Morad M. Functional coupling of Ca^{2+} channels and ryanodine receptors in cardiac myocytes. Proc Natl Acad Sci USA 1995; 92:121-125.
90. Means AR, VanBerkum MF, Bagchi I et al. Regulatory functions of calmodulin. Pharmacol Therapeut 1991; 50:255-270.
91. Gruver CL, DeMayo F, Goldstein MA et al. Targeted developmental overexpression of calmodulin induces proliferative and hypertrophic growth of cardiomyocytes in transgenic mice. Endocrinology 1993; 133:376-388.
92. Braun AP, Schulman H. The multifunctional calcium/calmodulin-dependent protein kinase: From form to function. Annu Rev Physiol 1995; 57:417-445.
93. Klee CB, Ren H, Wang X. Regulation of the calmodulin-stimulated protein phosphatase, calcineurin. J Biol Chem 1998; 273:13367-13370.
94. Lu J, McKinsey TA, Nicol RL et al. Signal-dependent activation of the MEF2 transcription factor by dissociation from histone deacetylases. Proc Natl Acad Sci USA 2000; 97:4070-4075.
95. Muller J, Nemoto S, Laser M et al. Calcineurin inhibition and cardiac hypertrophy. Science 1998; 282:1007.
96. Luo Z, Shyu KG, Gualberto A et al. Calcineurin inhibitors and cardiac hypertrophy. Nat Med 1998; 4:1092-1093.
97. Zhang W, Kowal RC, Rusnak F et al. Failure of calcineurin inhibitors to prevent pressure overload left ventricular hypertrophy in rats. Circ Res 1999; 84:722-728.
98. Ding B, Price RL, Borg TK et al. Pressure overload induces severe hypertrophy in mice treated with cyclosporine, an inhibitor of calcineurin. Circ Res 1999; 84:729-734.

99. Zou Y, Hiroi Y, Uozumi H et al. Calcineurin plays a critical role in the development of pressure overload-induced cardiac hypertrophy. Circulation 2001; 104:97-101.
100. De Windt LJ, Lim HW, Bueno OF et al. Targeted inhibition of calcineurin attenuates cardiac hypertrophy in vivo. Proc Natl Acad Sci USA 2001; 98:3322-3327.
101. Rothermel BA, McKinsey TA, Vega R et al. Myocyte-enriched calcineurin-interacting protein, MCIP1, inhibits cardiac hypertrophy in vivo. Proc Natl Acad Sci USA 2001; 98:328-333.
102. Sussman MA, Lim HW, Gude N et al. Prevention of cardiac hypertrophy in mice by calcineurin inhibition. Science 1998; 281:1690-1693.
103. Meguro T, Hong C, Asai K et al. Cyclosporine attenuates pressure overload hypertrophy in mice while enhancing susceptibility to decompensation and heart failure. Circ Res 1999; 84:735-740.
104. Shimoyama M, Hayashi D, Takimoto E et al. Calcineurin plays a critical role in pressure overload-induced cardiac hypertrophy. Circulation 1999; 100:2449-2454.
105. Lim HW, De Windt LJ, Steinberg L et al. Calcineurin expression, activation, and function in cardiac pressure overload hypertrophy. Circulation 2000; 101:2431-2437.
106. Hill JA, Karimi M, Kutschke W et al. Cardiac hypertrophy is not a required compensatory response to short-term pressure overload. Circulation 2000; 101:2863-2869.
107. Molkentin JD, Lu JR, Antos CL et al. A calcineurin-dependent transcriptional pathway for cardiac hypertrophy. Cell 1998; 93:215-228.
108. De Windt L, Lim H, Taigen T et al. Calcineurin-mediated hypertrophy protects cardiomyocytes from apoptosis in vitro and in vivo: An apoptosis-independent model of dilated heart failure. Circ Res 2000; 86:255-263.
109. Shibasaki F, McKeon F. Calcineurin functions in Ca^{2+}-activated cell death in mammalian cells. J Cell Biol 1995; 131:735-743.
110. Brewster JL, de Valoir T, Dwyer ND et al. An osmosensing signal transduction pathway in yeast. Science 1993; 259:1760-1763.
111. De Windt LJ, Lim HW, Haq S et al. Calcineurin promotes protein kinase C and c-Jun NH2-terminal kinase activation in the heart. Cross-talk between cardiac hypertrophic signaling pathways. J Biol Chem 2000; 275:13571-13579.
112. Wang HG, Pathan N, Ethell IM et al. Ca^{2+}-induced apoptosis through calcineurin dephosphorylation of BAD. Science 1999; 284:339-343.
113. Saito S, Hiroi Y, Zou Y et al. β-Adrenergic pathway induces apoptosis through calcineurin activation in cardiac myocytes. J Biol Chem 2000; 275:34528-34533.
114. Lotem J, Kama R, Sachs L. Suppression or induction of apoptosis by opposing pathways downstream from calcium-activated calcineurin. Proc Natl Acad Sci USA 1999; 96:12016-12020.
115. Olivetti G, Abbi R, Quaini F et al. Apoptosis in the failing human heart. N Engl J Med 1997; 336:1131-1141.

CHAPTER 7

The Role of Adrenoceptors in Mechanotransduction

Klaus-Dieter Schlüter,* Hans Michael Piper and Sibylle Wenzel

Abstract

Adrenoceptors are a large family of seven membrane spanning G-protein coupled receptors involved in many regulatory processes of the heart. Under conditions of mechanical load to heart, i.e., pressure overload, an activation of the sympathetic nerve system leads directly to stimulation of receptors of this family. Especially α-adrenoceptors are constantly coupled to regulation of protein synthesis and their stimulation leads to an imbalance of protein synthesis and degradation causing myocardial hypertrophy. Moreover, events initially evoked by coactivation of the renin-angiotensin-system, including the activation of cytokines like TGF-β, are able to induce an additional coupling of β₂-adrenoceptors to the regulation of protein synthesis, further favouring an imbalance of protein synthesis and degradation. Thus, several adrenoceptors are involved in a complex network of external and internal signals, finally leading to an adaptive response of the heart to mechanical load. The present review summarises our current understanding of these signal transduction pathways and their contribution to myocardial hypertrophy and heart failure.

Introduction

Mechanical load is one of the main triggers of cardiac growth. For example, systolic blood pressure developed by the immature right and left ventricle at birth is approximately 20-25 mm Hg but rapidly increases in the left ventricle to reach adult levels of over 120 mm Hg by several weeks of age. Right ventricular pressure changes very little after birth. In adults, cardiac mass in the left ventricle caused by hypertrophic growth of neonatal cardiomyocytes largely exceeds that of the right ventricle. In general, this mechanical dependent adaptation of heart mass in the left ventricle is a compensatory response of the heart. Due to the Laplace equation, increased ventricular mass reduces wall tension and therefore allows the heart to work in an appropriate mode. Nevertheless, cardiac hypertrophy in adults, as expressed as higher left ventricular mass compared to normal subjects, is an independent risk factor, associated with increased mortality and morbidity. Thus, although the adult heart is still able to compensate in case of increased afterload by increased cardiac mass, this hypertrophic response is unstable and subsequently leads to heart failure. The mechanisms involved in physiological and pathophysiological adaptation of ventricular mass to increased blood pressure, named pressure induced hypertrophy, are not fully understood. However, irrespectively of the fact that mechanical load by itself can induce signals normally involved in myocardial hypertrophy, many studies have shown that activation of adrenoceptors accompanies cardiac

*Corresponding Author: Klaus-Dieter Schlüter—Physiologisches Institut, Aulweg 129, 35392 Giessen, Germany.

Cardiac Mechanotransduction, edited by Matti Weckström and Pasi Tavi.
©2007 Landes Bioscience and Springer Science+Business Media.

hypertrophy induced by mechanical overload. In addition, in pressure overloaded hearts lowering of blood pressure directly by hydralazine was found to be insufficient to reduce hypertrophy and heart failure, but inhibition of adrenoceptors was sufficient. In line with these suggestions transgenic mice lacking dopamine-hydroxylase, void of the ability to synthesize norepinephrine, do not generate myocardial hypertrophy in response to pressure overload by transverse aortic constriction.[1] This indicates that catecholamines are playing a key role in the transduction process leading from mechanical load to hypertrophic growth.

Adrenoceptors in the Heart

Catecholamines act via stimulation of adrenoceptors, which are widely expressed on cardiac cells. They can be subdivided in several classes of receptors, i.e., in α- and β-adrenoceptors. α-Adrenoceptors are further subdivided into α_1- and α_2-adrenoceptors. α_1-Adrenoceptors are located on ventricular cardiomyocytes and thereby able to modify growth and activity of the cells directly. Three different α_1-adrenoceptors are known on the basis of pharmacological and molecular analysis and named α_{1A}-, α_{1B}-, and α_{1D}-adrenoceptors.[2] At least in the rat heart, expression of mRNA of all three subtypes has been shown.[3] Expression of mRNA and binding studies in the rat heart have revealed that the α_{1B} subtype is the dominant receptor subtype in the heart. Although in other mammalians than the rat the total expression of α_1-adrenoceptors in the myocardium is less pronounced, expression of at least α_{1A}- and α_{1B}-adrenoceptors has been demonstrated in all species investigated so far, including man. In contrast, α_2-adrenoceptors are not expressed on cardiomyocytes but involved in the regulation of expression and release of catecholamines from the transmural nerve endings of the heart. β-Adrenoceptors are also located in ventricular cardiomyocytes and they are, like the α-adrenoceptors, able to modify growth and contractile activity of cardiomyocytes. β-Adrenoceptors are also subdivided into several subtypes, called β_1-, β_2-, and β_3-adrenoceptors. Expression of all three receptor subtypes on the mRNA level and their functional coupling on cardiomyocytes has been shown. Figure 1 summarizes the cardiac expression of adrenoceptors. The β_1-adrenoceptor subtype seems to be the dominant isoform. Expression and functional coupling of adrenoceptors are targets for modification during the development of heart failure and/or hypertrophy. Stimulation of receptors causes downregulation of the receptor or signals distinct from the receptor level. Signals induced by stimulation of adrenoceptors can also interfere with each other, i.e., α-and β-adrenoceptors influence each other's activity as explained later. Taken together, the different expression of several adrenergic receptor isoforms, the modification of receptor expression under physiological and pathophysiological conditions, and the interaction of intracellular signals induced by each of these agonists strongly suggest a complex network of signals involved in the regulation of cardiac size and function under basal conditions and mechanical load.

The Role of α_1-Adrenoceptors in Pressure Induced Hypertrophy

In vivo adaptation of heart muscle mass to load is achieved either directly by mechanical forces to which the cells are exposed or indirectly via an activation of the sympathetic nerve system. Load adaptation to increased afterload leads to a release of catecholamines, predominantly norepinephrine, from transmural nerve endings of the heart (reviewed in ref. 4). The main transmitter of the sympathetic nerve system, norepinephrine, stimulates myocardial growth directly. Experiments in vivo and ex vivo indicate that norepinephrine increases protein synthesis. This hypertrophic growth response to norepinephrine is evoked at least in part via stimulation of α_1-adrenoceptors.[5-6] On isolated neonatal rat cardiomyocytes, which were allowed to beat spontaneously, stimulation of α_1-adrenoceptors is sufficient to induce myocardial hypertrophy.[7] Moreover, the phenotype induced by α_1-adrenoceptor dependent myocardial hypertrophy mimics all aspects of myocardial hypertrophy including reexpression of foetal type proteins like atrial natriuretic factor, β-myosin heavy chain, α-smooth muscle actin, and creatine kinase B.[8-11] Historically, α_1-adrenoceptor stimulation was the first receptor signal system found to influence directly expression of cardiac genes which are normally induced in pressure

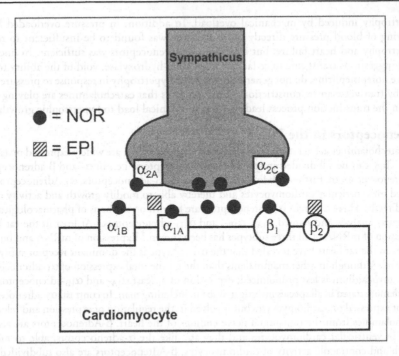

Figure 1. Distribution of adrenergic receptors in the ventricle and the corresponding neurotransmitters. Activation of the sympathetic nerve system as a response of mechanical load of the heart leads to a release from predominantly norepinephrine (filled circles) and epinephrine (dashed boxes) to a minor extent. Norepinephrine binds to all receptors except the β_2-adrenoceptor. Cardiomyocytes express two types of α-adrenoceptors and β-adrenoceptors and their relative expression is indicated by the size of symbols. Not shown are α_{1D}- and β_3-adrenoceptors which participation in load dependent growth control is poorly characterized and described. α_2-adrenoceptors are located on the transmural nerve endings.

induced hypertrophy. In subsequent studies on adult rat ventricular cardiomyocytes, stimulation of α_1-adrenoceptors was found to be a robust and strong signal in regard to its ability to induce hypertrophy.[11] The coupling between α-adrenoceptor stimulation and the regulation of protein synthesis was not influenced in any way by culturing the cells for several days irrespectively from morphological changes in these cultures.[12-13] Indeed, no stronger hypertrophic stimulus has been found than direct stimulation of α_1-adrenoceptors in a given in vitro assay when compared directly. α-Adrenoceptor-dependent induction of hypertrophic growth as part of the mechanosensitive adaptation of the heart has been used to characterise intracellular signal transduction steps leading to myocardial hypertrophy.

α_1-Adrenoceptors and Their Involvement in Pressure Induced Hypertrophy

One question under investigation is which α_1-adrenoceptor subtype triggers the hypertrophic response of norepinephrine. There is good evidence from experiments on neonatal and adult rat cardiomyocytes that α_{1A}-adrenoceptor stimulation directly causes hypertrophy, while stimulation of α_{1B}-adrenoceptors does not.[14-15] This leads to the undissolved question what specifically the role of the dominant α-adrenoceptor subtype, namely α_{1B} is. Coupling of α_{1A}- and α_{1B}-adrenoceptors to phospholipase C indicates the functional relevance of both receptor subtypes.[15] α_{1A}- but not α_{1B}-adrenoceptor stimulation, activates phospholipase D as well, indicating functional differences between the receptors.[16] Exposure of cardiomyocytes to norepinephrine does not only activate different α-adrenoceptors but also regulate their expression. Norepinephrine downregulates the α_{1B}-adrenoceptor mRNA leading to an improved α_{1A}/α_{1B} ratio in neonatal cardiomyocytes from rats but similar changes were found in their expression pattern in vivo under pressure overload.[17] Neonatal cardiomyocytes show an antagonistic relationship between α_{1B}-adrenoceptors and α_{1A}-adrenoceptors, which might contribute to the functional coupling of α_{1A}-adrenoceptors.[18] When all α-adrenoceptor subtypes on cardiomyocytes were knocked out in transgenic mice, banding of the aorta did no longer induce hypertrophy, indicating participation of α-adrenoceptor stimulation in pressureinduced hypertrophy from mice.[19] Transgenic mice with myocardial overexpression of wild type α_{1B}-adrenoceptors develop severe myocardial hypertrophy when the mice were treated with phenylephrine, an α-adrenoceptor agonist.[20] Overexpression of the α_{1B}-adrenoceptor induces left ventricular dysfunction but not hypertrophy.[21] Myocardial expression of a constitutively active α_{1B}-adrenergic receptor in transgenic mice produces cardiac hypertrophy.[22] Exposed to thoracic aortic banding mice lacking the α_{1B}-adrenoceptor develop hypertrophy and show no cardiovascular remodelling provoked by a subpressor dosis of phenylephrine.[23] Targeted α_{1A}-adrenergic receptor overexpression induces enhanced cardiac contractility but not hypertrophy.[24] In vitro, α-adrenoceptor stimulation in mice myocytes is not sufficient to induce regular intracellular signalling.[25] Thus, although extended work on transgenic animals in regard to α-adrenoceptor stimulation was performed it is not clear which specific role this type of receptors plays in mice. α-Adrenoceptor expression in rats exceeds that of other mammalian species including mice by a factor of 5-7. In summary, mice differ from rat in regard to the contribution of α-adrenoceptor stimulation and their response to pressure overload, which limits the use of transgenic mice. Most of the results on rats supported so far indicate a dominant participation of the α_{1A}-adrenoceptor in this process.

In spontaneously hypertensive rats and cardiomyopathic Syrian hamster stimulation of α_1-adrenoceptors is part of pressure induced myocardial hypertrophy.[26-27] The expression of α_1-adrenoceptors increases in the early phase of hypertension and the sensitivity of the receptors increases at the prehypertensive state.[26,28] Aortic banding induces pressure overload and causes an early increase in the number of α-adrenoceptors in normotensive rats.[29] Conversely, even a sub-pressure dose of the α-adrenoceptor antagonist bunazosin was able to reduce myocardial hypertrophy in rats, indicating that activation of the sympathetic nerve system via stimulation of α-adrenoceptors causes myocardial hypertrophy rather than the mechanical load itself.[30] It should be pointed out that the contribution of α-adrenoceptors to the development of pressure induced hypertrophy found in the rat could be confirmed in man. In clinical studies bunazosin, an α-adrenoceptor antagonist, significantly reduced left ventricular mass in hypertensive patients.[31-32] However, inhibition of α-adrenoceptors as treatment of hypertensive patients to reduce myocardial hypertrophy is limited by side effects of α-inhibitors.

The Signalling of α_1-Adrenoceptors in Pressure Induced Hypertrophy

α-Adrenoceptors belong to a family of seven transmembrane spanning domain receptors coupled to $G\alpha_q$. The importance of $G\alpha_q$ in mediating the hypertrophic effect was shown directly by generating transgenic mice overexpressing wild type $G\alpha_q$.[33] At a threshold of 4-fold overexpression these mice develop cardiac hypertrophy with all characteristics for hypertrophy and heart failure. On the other hand, transgenic mice lacking G_q are unable to induce hypertrophy in presence of pressure overload.[34] Once α_1-adrenoceptors are activated by binding of a ligand, they activate phospholipase C (PLC) in a $G\alpha_q$-dependent way. α_1-Adrenoceptors, G_q and phospholipase C are found in the same caveolin fraction.[35] This strongly suggests a coupling of these three molecules. PLC generates IP3 and diacylglycerol. Another source for diacylglycerol is an activation of phospholipase D (PLD) via stimulation of α_{1A}-adrenoceptors.[16] Diacylglycerol activates membrane bound protein kinase C (PKC). Work on transgenic animals has provided evidence that PKC activation is sufficient to induce hypertrophic growth.[36] Further evidence for a causal role of PKC activation in triggering the hypertrophic response

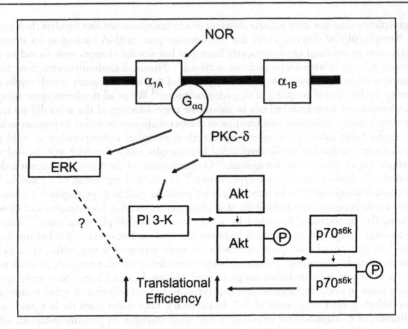

Figure 2. Increase in translational efficiency by norepinephrine in rat cardiomyocytes. Norepinephrine induces protein synthesis via stimulation of α_{1A}-adrenoceptors. The figure summarizes those signal transduction steps which have been constantly found to play a role by different groups. These are the activation of protein kinase C (PKC), phosphatidylinositiol 3-kinase (PI 3-K), phosphorylation of the Akt-kinase (Akt), and finally phosphorylation of p70[s6k] which leads to a phosphorylation of ribosomal proteins. The integration of further signal transduction steps in this flow chart is not clear as mentioned in the text. Activation of the ERK pathway might contribute to the increase in translational activity, but this was not confirmed by all groups and the exact mechanism is not clear as well.

comes from work on transgenic mice with cardiac specific overexpression of $G\alpha_q$ which showed an activation of PKC-ε. More directly, participation of PKC activation in the hypertrophic response evoked by α-adrenoceptor stimulation comes from experiments in which PKC activation was blocked by pharmacological treatment of cardiomyocytes. Direct stimulation of protein kinase C by phorbol ester is sufficient to induce hypertrophic growth.[37] α-Adrenoceptor stimulation causes an activation of a bride spectrum of PKC isoforms, from which the activation of PKC-δ seemed to be of particulate interest in regard to growth control.[38-39]

The signals distinct from PKC activation must be subdivided into signals leading to an increased protein synthesis (hypertrophic growth) and signals involved in induction of transcription of specific genes (phenotype changes). There is a large body of evidence that the acceleration of protein synthesis by α-adrenoceptor stimulation is evoked by a PKC-dependent activation of the PI 3-kinase/Akt pathway leading to a downstream activation of p70[s6k].[40-42] Figure 2, summarises signal transduction steps identified in the α_{1A}-adrenoceptor pathway leading to hypertrophy. PI 3-kinase/Akt activation is not specific for the hypertrophic effect evoked by α-adrenoceptor stimulation and commonly used by other hypertrophic stimuli as well. Overexpression of a constantly active form of Akt results in a hypertrophic phenotype similar to that evoked by α-adrenoceptor stimulation.[43] This increase in translational activity is accompanied by an increase in translational capacity, as indicated by an increase in total RNA, which mainly reflects ribosomal RNA.[5] An increase in RNA synthesis[44,12] and activation of transcription factors for ribosomal RNA[45] further underline the increase in translational capacity.

There is an ongoing discussion in the literature whether or not an activation of the early response kinase (ERK) -pathway contributes to the increase in protein synthesis. On neonatal cardiomyocytes evidence for participation of ERK activation in the hypertrophic response by α-adrenoceptor stimulation has been suggested but not constantly found.[46-47] Similar, on adult ventricular cardiomyocytes, most investigators found no evidence for a contribution of the ERK pathway in the regulation of protein synthesis caused by α-adrenoceptor stimulation.[48,15] There are some exceptions.[49,50] All investigators could demonstrate, however, an activation of the ERK pathway by α-adrenoceptor stimulation. The question whether ERK activation does play a role in α-adrenoceptor-dependent regulation of cardiac growth or not can therefore not be finally answered. Critical evaluation of studies performed so far with quite contradictory results justifies the common conclusion that an activation of the ERK pathway can contribute to the regulation of protein synthesis under specific conditions, which have not been characterised so far (see Fig. 2). Other members of the mitogen-activated kinase family, i.e., p38 and c-jun kinase (JNK), seem not to be activated by α-adrencoeptor-dependent stimulation.[50]

There are numerous open questions in regard to the signal transduction process of α-adrenoceptor stimulation. This includes the observation that further signals, i.e., activation of sodium-proton-exchanger (NHE)-1, phospholipase D, some G-proteins like Rlf, reactive oxygen species, and voltage dependent calcium channels are causally involved in the hypertrophic response. Each of these signals can be activated by α-adrenoceptor stimulation.[16,37,51-53] It is unclear at present how these signals integrate in the above mentioned pathways and these questions need further clarification.

In summary, stimulation of α-adrenoceptors contributes significantly to the hypertrophic response of a mechanical loaded heart as shown on rats, mice, hamsters, and even in man. The involvement of α_{1A}- and α_{1B}-adrenoceptors might be different in mice and rats. In any way, stimulation of α_1-adrenoceptors activates PKC and PKC-dependent signal transduction pathways which interfere directly with increases in translational activity, capacity, and changes in gene expression as normally found in hypertrophic hearts. The participation of the ERK-pathway is uncertain and needs further clarification identifying the conditions under which this pathway contributes to hypertrophic growth.

The Role of α_2-Adrenoceptors in Pressure Induced Hypertrophy

Stimulation of α_2-adrenoceptors does not play a direct role in cardiac hypertrophy as these receptors are not expressed on cardiomyocytes. However, α_2-adrenceptors are indirectly involved in the mechano-induced growth control of the heart as they control the release of catecholamines to which the heart is exposed. From three different known α_2-adrenoceptors, namely A, B and C, two are involved in the modulation of neurotransmitter release (see Fig. 3). These are the subtypes A and C. Both of them have a different role in the regulation of neurotransmitter release depending from the heart rate.[54] Cardiac pressure overload by transverse aortic constriction in mice lacking one of the two different α_2-adrenoceptors developed severe hypertrophy and significantly reduced survival rates compared to either wild type or α_{2B}-adrenoceptor deficient mice. Changes in the functional coupling or expression of these presynaptic inhibitory autoreceptors are not known under pathophysiological conditions, however, patients with a dysfunctional variant of the α_{2C}-adrenoceptor developed severe heart failure.[55] These data indicate again that a constant sympathetic overdrive is sufficient to mimic the load adaptation of the heart.

The Role of β_1-Adrenoceptors in Pressure Induced Hypertrophy

In contrast to the stimulation of α-adrenoceptors, which gave nearly identical results in vivo and in vitro, stimulation of β-adrenoceptors gave different results in vivo and in vitro. The most likely explanation for this different is, that β-adrenoceptor stimulation in vivo changes the hemodynamic properties, while on quiescent cells, β-adrenoceptor stimulation does not interfere with hemodynamic changes. Following these suggestions, it was directly

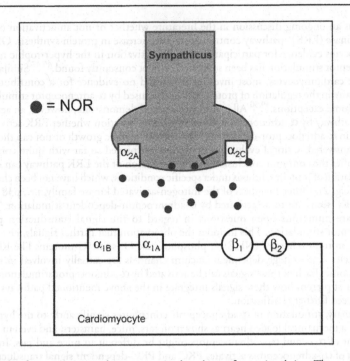

Figure 3. Role of presynaptic α_2-adrenoceptors in the release of norepinephrine. These receptors are occupied by norepinephrine and lead to an inhibition of norepinephrine release. Lack of functional active α_2-adrenoceptors leads therefore to a sympathetic overdrive.

demonstrated on neonatal cardiomyocytes that norepinephrine increases protein synthesis and thereby induces myocardial hypertrophy via its effect on cell contraction.[7] However, most of the experiments performed on the role of β-adrenoceptor stimulation in the regulation of myocardial hypertrophy were investigated by nonselective agonists like isoprenaline. Most recently it turned out that selective stimulation of β_1-adrenoceptors by itself is sufficient to increase protein synthesis moderately.[56] The physiological relevance of this observation is, however, not clear. The physiological agonist for β_1-adrenoceptors, norepinephrine increases protein synthesis more effectively via direct stimulation of α-adrenoceptors. Moreover, combined stimulation of α-and β-adrenoceptors, normally equally activated by norepinephrine, leads to an inhibition of the α-adrenoceptors via β-adrenoceptor stimulation (see Fig. 4).[57] Thus it might be concluded that under normal and physiological conditions, the impact of norepinephrine on protein synthesis is small, due to the cAMP-dependent reduction in protein kinase C activation. In chronic heart failure, however, expression of β_1-adrenoceptors declines leading to a more efficient coupling of α-adrenoceptor stimulation to protein synthesis.

Accordingly to the pure characterization of β_1-selective induced myocardial hypertrophy, there is little understanding about the intracellular signal transduction pathways. Two things seem to be remarkable. First, direct hypertrophic stimulation of β_1-adrenoceptors activates the classical adenylate cyclase pathway, but the hypertrophic response seems to be cAMP-independent.[56] Second, activation of p38 MAP kinase seems to be sufficient to induce hypertrophic growth under specific conditions, but this seems not to be sufficient in case of β_1-adrenoceptor stimulation.[25,58] As mentioned above the contribution of the ERK pathway in myocardial hypertrophy is still a matter of debate. However, activation of adenylate cyclase by selective stimulation of β_1-adrenoceptors attenuates ERK-activation by α-adrenoceptor

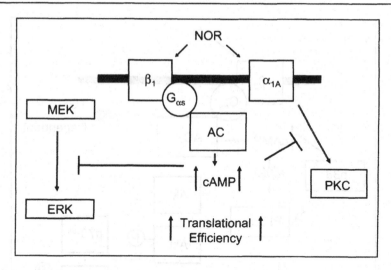

Figure 4. Interaction of β_1- and α_1-adrenoceptors on ventricular cardiomyocytes. Both receptors are occupied by norepinephrine at the same time. Stimulation of β_1-adrencoeptors leads to a cAMP-dependent inhibition of PKC-activation and ERK-activation by α-adrencoceptor stimulation and attenuates the hypertrophic response normally found by selective stimulation of α-adrenoceptors.

stimulation.[57,59] This indicates that myocardial hypertrophy by selective stimulation of β_1-adrenoceptors is independent from the ERK pathway.

In contrast to the missing details from extensive in vitro studies, there are several reports that confirm in vivo that specific induction of β_1-adrenoceptors induces myocardial hypertrophy. This includes work on transgenic mice overexpressing β_1-adrenoceptors.[60,61] Similarly, autoantibodies directed against β_1-adrenoceptors from patients with heart failure are sufficient to induce hypertrophic growth.[62] Transgenic mice overexpressing β_1-adrenoceptors develop myocardial hypertrophy and this is attenuated by inhibition of NHE-1.[61] As mentioned above, NHE-1 contributes to signalling pathways leading to myocardial hypertrophy of the α-adrenoceptor as well. Its precise role in the interaction of other well described pathways has to be established. However, it seems to be pivotal in different signal transduction pathways.

It is questionable whether stimulation of β_1-adrenoceptors contributes to the initiation of myocardial hypertrophy due to pressure overload. However, changes in the β_1-adrenoceptor system are clearly part of the scenario leading to a functional loss of cardiac function. β_1-adrenoceptors and their subsequent signal transduction elements are targets undergoing desensitisation during pressure overload of the heart. This desensitisation requires downregulation of the β_1-adrenoceptors.[63-65] Up-regulation of β-adrenoceptor kinase (β-ARK), which phosphorylates and desensitises β_1-adrenoceptors, and up-regulation of G_i-proteins also contributes to loss of functional β-adrenoceptors.[66] A causal relationship between β-ARK-dependent changes in the β_1-adrenoceptor signalling and heart failure due to pressure overload is established by experiments in which inhibition of β-ARK attenuates the progression of hypertrophy and failure.[67] In conclusion, stimulation of β_1-adrenoceptors seems not to be the main factor triggering pressureinduced hypertrophy via the sympathetic nerve system, however, they are involved in the transition from hypertrophy to heart failure.

The Role of β_2-Adrenoceptors in Pressure Induced Hypertrophy

β_2-adrenoceptors are less pronounced expressed on myocardial cells than β_1-adrenoceptors. Nevertheless, specifically β_2-adrenoceptors play a very interesting and specific role in the formation of myocardial hypertrophy under chronic pressure overload. A couple of years ago it

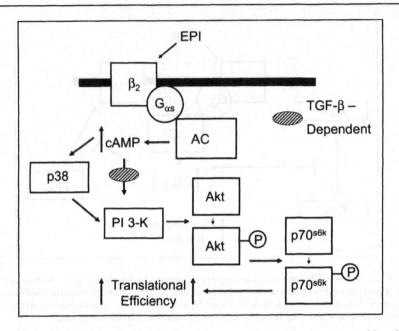

Figure 5. Signal transduction pathways leading to a hypertrophic growth response caused by selective stimulation of β_2-adrenoceptors via epinephrine. The initial steps are specific for this receptor, i.e., activation of adenylate cyclase (AC) and cAMP-dependent p38 MAP kinase, but end in the stimulation of further signal transduction pathways which are in common with other types of adrenoceptor-dependent stimulation of protein synthesis, i.e., caused by α-adrenoceptor stimulation (see also Fig. 2). Note that p38 MAP kinase activation is not sufficient for the induction of protein synthesis and needs further stimulation of signal transduction steps which require a TGF-β-dependent step.

was shown by work on isolated cardiomyocytes that isoprenaline is sufficient to increase protein synthesis on quiescent cardiomyocytes under specific conditions.[12,68] It turned out, that β_2-adrenoceptors are responsible for this effect.[69] The result is remarkable because it indicates that in addition to receptors constantly coupled to the regulation of protein synthesis, i.e., α-adrenoceptors, further receptor systems can imbalance the ratio of protein synthesis and degradation and thus contribute to myocardial hypertrophy. The signal transduction pathway involved in this specific effect of β_2-adrenoceptor-dependent myocardial hypertrophy has been characterised as well. It includes both signal transduction elements specific for β_2-adrenoceptors and those in common with what we have already learned from α-adrenoceptor stimulation. This induced signalling pathway depends on an activation of adenylate cyclase, p38 MAP kinase, PI 3 kinase, and p70[s6k] (see Fig. 5).[42,48,70,71] The activation of this pathway is not only hypertrophic but also anti-apoptotic.

Induction of translational efficiency is realised by the same mechanism than under α-adrenoceptor stimulation, namely phosphorylation of ribosomal proteins. In contrast, induction of translational capacity is caused via a complete different signal transduction pathway than that described above for α-adrenoceptors. It depends on an cAMP-dependent activation of ornithine decarboxylase (ODC), the rate limiting enzyme of polyamine metabolism, via de novo transcription of ODC.[72] The increase in polyamines stabilises preexisting rRNA and via reduction of RNA degradation increases the total amount of ribosomes (see Fig. 6). Induction of ODC by β-adrenoceptor stimulation was confirmed in vivo.[73] This is a specific event in all forms of β-adrenoceptor-dependent hypertrophy found so far. Participation of β_2-adrenoceptor stimulation in pressureinduced myocardial hypertrophy is established in transgenic animals overexpressing β_2-adrenoceptors. These mice develop hypertrophy especially in presence of

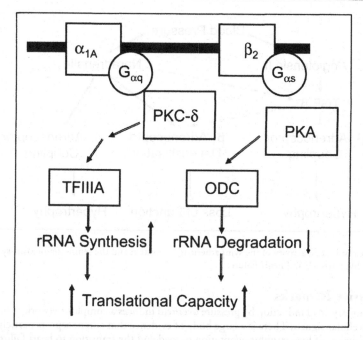

Figure 6. Costimulation of translational capacity by α- and β-adrenoceptor dependent pathways which increase RNA synthesis and decrease RNA degradation.

pressure overload by aortic banding.[74] Rats receiving the β₂-adrenoceptor specific agonist clenbuterol also develop myocardial hypertrophy especially under conditions of pressure overload.[75,76] However, they show no loss of function. This might indicate that stimulation of β₂-adrenoceptors by itself causes a type of myocardial hypertrophy not leading to a loss of function while stimulation of β₁-adrenoceptors, although less important for hypertrophy, seems to be more important in triggering malfunction.

The question via which mechanism β₂-adrenoceptors are induced to couple to the regulation of protein synthesis is only partially analysed. However, the main steps identified so far indicate that key steps involved in the induction of such a hypertrophic coupling of β₂-adrenoceptors are strongly activated as a response to mechanical load of the heart. First, it was found that the induction of hypertrophic coupling of β₂-adrenoceptors is mediated via TGF-β.[77] This cytokine is expressed in cardiomyocytes itself and regulated in its expression by β-adrenoceptor stimulation.[78] Once TGF-β is released and activated it induces a coupling of the β₂-adrenoceptor system to the regulation of protein synthesis. Most likely, the critical step is located between the cAMP-dependent activation of the stress-activated MAP kinase p38 and the subsequent activation of PI 3-kinase. The mechanism by which TGF-β induces this coupling is not clear. However, induction of TGF-β as indicated by transient accelerated expression is commonly found in hypertensive rats and, remarkable enough, this happens at the transition from myocardial hypertrophy to heart failure. Thus, one of the key questions is to identify the trigger for the induction of TGF-β. There is some evidence that this is mediated via the renin-angiotensin-system, especially angiotensin II, and the cellular signals involved in this process have already been started to be analysed.[79] A strong interaction between the renin-agiotensin-system and the sympathetic nerve system in regard to mechanosensitive growth control of the heart has been demonstrated in several studies, indicating that inhibition of one of these pathways interferes with the other one. TGF-β seems to be one of the links between these main pathways.

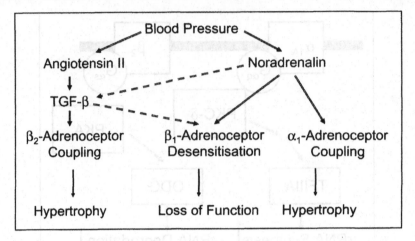

Figure 7. Coordinated activity of the sympathetic nerve system and the renin-angiotensin-system leads to cardiac hypertrophy and heart failure.

Conclusive Remarks

In summary, load induction by pressure overload induces a complex network of both extra- and intracellular signals. There is a large body of evidence that adrenoceptors participate both in the induction of hypertrophic adaptation to load and the transition to heart failure as well. These pathways should not be considered as signal transduction pathways in addition to other pathways but in relation to them. There are two different networks: The first comprises the interaction of α_1- and β_1-adrenoceptors and limits the hypertrophic response of the heart to norepinephrine under physiological conditions. The second comprises an interaction of the renin-angiotensin-system, locally produced cytokines like TGF-β, and finally the sympathetic nerve system itself. The latter one contributes to load adaptation of the heart (see Fig. 7).

References

1. Rapaccioulo A, Esposito G, Caron K et al. Important role of endogenous norepinpehrine and epinephrine in the development on in vivo pressure overload cardiac hypertrophy. J Am Coll Cardiol 2001; 38:876-882.
2. Michel MC, Kenny B, Schwinn DA. Classification of α_1-adrenoceptor subtypes. Naunyn-Schmiedeberg's Arch Pharmacol 1995; 352:1-10.
3. Price DT, Chari RS, Berkowitz DE et al. Expression of α_1-adrenergic receptor subtype mRNA in rat tissues and human SK-N-MC neuronal cells: Implications for α_1-adrenergic receptor subclassification. Mol Pharmacol 1994; 46:221-226.
4. Morgan HE, Baker KM. Cardiac hypertrophy mechanical, neural and endocrine dependence. Circulation 1991; 83:13-25.
5. Zierhut W, Zimmer HG. Significance of myocardial α- and β-adrenoceptors in catecholamine-induced cardiac hypertrophy. Circ Res 1989; 65:14717-1425.
6. Fuller SJ, Gaitanaki CJ, Sugden PH. Effects of catecholamines on protein synthesis in cardiac myocytes and perfused hearts isolated from adult rats. Biochem J 1990; 266:727-736.
7. Simpson PC. Stimulation of hypertrophy of cultures neonatal rat hearts cells through an α_1-adrenergic receptor and induction of beating through an α_1- and β_1-adrenergic receptor interaction: Evidence for independent regulation of growth and beating. Circ Res 1985; 56:884-894.
8. Ardati A, Nemer M. A nuclear pathway for α_1-adrenergic receptor signalling in cardiac cells. EMBO J 1993; 12:5131-5139.
9. Waspe LE, Ordahl CP, Simpson PC. The cardiac β-myosin heavy chain isogene is induced selectively in α_1-adrenergic receptor-stimulated hypertrophy of cultured rat heart myocytes. J Clin Invest 1990; 85:1206-1214.
10. Long CS, Ordahl CP, Simpson PC. α_1-Adrenergic receptor stimulation of sarcomeric actin isogene transcription in hypertrophy of cultured rat heart muscle cells. J Clin Invest 1989; 83:1078-1082.

11. Schlüter K-D, Piper HM. Trophic effects of catecholamines and parathyroid hormone on adult ventricular cardiomyocytes. Am J Physiol Heart Circ Physiol 1992; 263:H1739-H1746.
12. Pinson A, Schlüter K-D, Zhou XJ et al. α- and β-Adrenergic stimulation of protein synthesis in cultured adult ventricular cardiomyocytes. J Mol Cell Cardiol 1993; 25:477-490.
13. Ikeda UY, Tsuruya Y, Yaginuma T. α_1-Adrenergic stimulation is coupled to cardiac myocyte hypertrophy. Am J Physiol Heart Circ Physiol 1991; 260:H953-H956.
14. Knowlton KU, Michel MC, Itani M et al. The α_1-adrenoceptor subtype mediates biochemical, molecular, and morphological features of cultured myocardial cell hypertrophy. J Biol Chem 1993; 268:15374-15380.
15. Pönicke K, Schlüter K-D, Heinroth-Hoffmann I et al. Noradrenalin-induced increase in protein synthesis in adult rat cardiomyocytes: Involvement of only α_{1A}-adrenoceptors. Naunyn-Schmiedeberg's Arch Pharmacol 2001; 364:444-453.
16. Mier K, Kemken D, Katus HA et al. Adrenergic activation of cardiac phospholipase D: Role of α_1-adrenoceptor subtypes. Cardiovasc Res 2002; 54:133-139.
17. Rokosh DG, Stewart AF, Chang KC et al. α_1-Adrenergic receptor subtype mRNAs are differentially regulated by α_1-adrenergic and other hypertrophic stimuli in cardiac myocytes in culture and in vivo. Repression of α_{1B} and α_{1D} but not of α_{1C}. J Biol Chem 1996; 271:5839-5843.
18. Deng XF, Sculptoreanu A, Mulay S et al. Crosstalk between α_{1A}- and α_{1B}-adrenoceptors in neonatal rat myocardium: Implications in cardiac hypertrophy. J Pharmacol Exptl Therap 1998; 286:489-196.
19. O'Connell TD, Stigwart PM, Simpson GL et al. α_1-Adrenergic receptors are required for stress response in the heart. Circulation 2002; 106(Suppl):II-58, [Abstract].
20. Iaccarino G, Keys JR, Rapacciuolo A et al. Regulation of myocardial βARK1 expression in catecholamine-induced cardiac hypertrophy in transgenic mice overexpressing α_{1B}-adrenergic receptors. J Am Coll Cardiol 2001; 38:534-540.
21. Grupp IL, Lorenz JN, Walsh RA et al. Overexpression of α_{1B}-adrenergic receptor induces left ventricular dysfunction in the absence of hypertrophy. Am J Physiol Heart Circ Physiol 1998; 275:H1338-H1350.
22. Milano CA, Dolber PC, Rockman HA et al. Myocardial expression of a constitutively active α_{1B}-adrenergic receptor in transgenic mice induces cardiac hypertrophy. Proc Natl Acad Sci USA 1994; 91:10109-10113.
23. Vecchione C, Fratta L, Tizzoni D et al. Cardiovascular influences of α_{1B}-adrenergic receptor deficient mice. Circ 2002; 105:1700-1707.
24. Lin F, Owens WA, Chen S et al. Targeted α_{1A}-adrenergic receptor overexpression induces enhanced cardiac contractility but not hypertrophy. Circ Res 2001; 89:343-350.
25. Sabri A, Pak E, Allcott SA et al. Coupling function of endogenous α_1- and β-adrenergic receptors in mouse cardiomyocytes. Circ Res 2000; 86:1047-1053.
26. Imai C, Tozawa M, Sunagawa O et al. Alterations in α_1-adrenergic receptor densities in right and left ventricles of spontaneously hypertensive rats. Biol Pharm Bull 1995; 18:1001-1005.
27. Kagiya T, Hori M, Iwakura K et al. Role of increased α_1-adrenergic activity in cardiomyopathic Syrian hamster. Am J Physiol Heart Circ Physiol 1991; 260:H80-H88.
28. Böhm M, Mende U, Schmitz U et al. Increased sensitivity to α-adrenoceptor stimulation but intact purinergic and muscarinergic effects in prehypertensive cardiac hypertrophy of spontaneously hypertensive rats. Naunyn-Schmiedeberg's Arch Pharmacol 1986; 333:284-289.
29. Ganguly PK, Lee SL, Beamish RE et al. Altered sympathetic system and adrenoceptors during the development of cardiac hypertrophy. Am Heart J 1989; 118:520-525.
30. Izumi Y, Matsuoka M, Kubo A et al. Prevention of cardiac hypertrophy by a sub-antihypertensive dose of the α_1-adrenergic antagonist bunazosin in Dahl salt-sensitive rats. Hypertens Res 1996; 19:147-150.
31. Schobel HP, Langenfeld M, Gatzka C et al. Treatment and post-treatment effects of α-versus β-receptor blockers on left ventricular structure and function in essential hypertension. Am Heart J 1996; 132:1004-1009.
32. Veelken R, Schmieder RE. Overview of α_1-adrenoceptor antagonism and recent advances in hypertensive therapy. Am J Hypertens 1996; 9:S139S-149S.
33. D'Angelo DD, Sakata Y, Lorenz JN et al. Transgenic $G\alpha_q$ overexpression induces cardiac contractile failure in mice. Proc Natl Acad Sci USA 1997; 94:8121-8126.
34. Wettschureck N, Rütten H, Zywietz A et al. Absence of pressure overload induced myocardial hypertrophy after conditional inactivation of $G\alpha_q/G\alpha_{11}$ in cardiomyocytes. Nat Med 2001; 7:1236-1240.

35. Fujita T, Toya Y, Iwatsubo K et al. Accumulation of molecules involved in α₁-adrenergic signal within caveolae: Caveolin expression and the development of cardiac hypertrophy. Cardiovasc Res 2001; 51:709-716.
36. Bowman JC, Steinberg SF, Jiang T. Expression of protein kinase C β in the heart causes hypertrophy in adult mice and sudden death in neonates. J Clin Invest 1997; 100:2189-2195.
37. Schäfer M, Schäfer C, Piper HM et al. Hypertrophic responsiveness of cardiomyocytes to α- and β-adrenoceptor stimulation requires sodium-proton-exchanger-1 (NHE-1) activation but not cellular alkalization. Eur J Heart Failure 2002; 4:249-254.
38. Ruf S, Piper HM, Schlüter K-D. Specific role for the extracellular signal-regulated kinase pathway in angiotensin II- but not in phenylephrine-induced cardiac hypertrophy in vitro. Pflügers Arch Eur J Physiol 2002; 443:483-490.
39. Rohde S, Sabri S, Kamasamudran R et al. The α₁-adrenoceptor subtype and protein kinase C isoform-dependence of norepinephrine's actions in cardiomyocytes. J Mol Cell Cardiol 2000; 32:1193-1209.
40. Sadoshima J, Qui Z, Morgan JP et al. Angiotensin II and other hypertrophic stimuli mediated by G protein-coupled receptors activate tyrosine kinase, mitogen-activated protein kinase, and 90- kD S6 kinase in cardiac myocytes. The critical role of Ca²⁺-dependent signalling. Circ Res 1995; 76:1-55.
41. Bolouyt-MO, Zheng JS, Younes A et al. Rapamycin inhibits α₁-adrenergic receptor-stimulated cardiac myocyte hypertrophy but not activation of hypertrophiy-associated genes. Evidence for involvement of p70 S6 kinase. Circ Res 1997; 81:176-182.
42. Schlüter K-D, Goldberg Y, Taimor G et al. Role of phosphatidylinositol 3-kinase activation in the hypertrophic growth of adult ventricular cardiomyocytes. Cardiovasc Res 1998; 40:174-181.
43. Matsui T, Li L, Wu JC et al. Phenotypic spectrum caused by transgenic overexpression of activated Akt in the heart. J Biol Chem 2002; 277:22896-22901.
44. McDermott PJ, Rothblum LI, Smith SD et al. Accelerated rates of ribososmal RNA synthesis during growth of contracting heart cells in culture. J Biol Chem 1989; 264:18220-18227.
45. Kanevskij M, Taimor G, Schäfer M. Neuropeptide Y modifies the hypertrophic response of adult ventricular cardiomyocytes to norepinephrine. Cardiovasc Res 2002; 53:879-887.
46. Glennon PE, Kaddoura S, Sale EM et al. Depletion of mitogen-activated protein kinase using antisense oligonucleotide approach downregulates the phenylephrine-induced hypertrophic response in rat cardiac myocytes. Circ Res 1996; 78:954-961.
47. Thorborn J, Frost JA, Thorburn A. Mitogen-activated protein kinase mediate changes in gene expression, but not cytoskeletal organization associated with cardiac muscle cell hypertrophy. J Cell Biol 1994; 126:1565-1572.
48. Schlüter K-D, Simm A, Schäfer M et al. Early response kinase and PI 3-kinase activation in adult cardiomyocytes and their role in hypertrophy. Am J Physiol Heart Circ Physiol 1999; 276:H1655-H1663.
49. Wang L, Proud CG. Ras/Erk signalling is essential for activation of protein synthesis by Gq protein-coupled receptor agonists in adult cardiomyocytes. Circ Res 2002; 91:821-829.
50. Xiao L, Pimental DR, Amin JK et al. MEK1/2-ERK1/2 mediates α₁-adrenergic receptor-stimulated hypertrophy in adult rat ventricular myocytes. J Mol Cell Cardiol 2001; 33:779-787.
51. Post GR, Swiderski C, Waldrop BA et al. Guanine nucleotide exchange factor-like factor (Rfl) induces gene expression and potentiates α₁-adrenergic receptor-induced transcriptional responses in neonatal rat ventricular myocytes. J Biol Chem 2002; 277:15286-15292.
52. Amin JK, Xiao L, Pimental DR et al. Reactive oxygen species mediate α-adrenergic receptor-stimulated hypertrophy in adult rat ventricular myocytes. J Mol Cell Cardiol 2001; 33:131-139.
53. Eble DM, Qi M, Waldschmidt S et al. Contractile activity is required for sarcomeric assembly in phenylephrine-induced cardiac myocyte hypertrophy. Am J Physiol 1998; 274:C1226-1237.
54. Hein L, Altman JD, Kobilka BK. Two functionally distinct α₂-adrenergic receptors regulate sympathetic neurotransmission. Nature 1999; 402:181-184.
55. Brede M, Wiesmann F, Jahns R et al. Feedback inhibition of catecholamine release by two different α₂-adrenoceptor subtypes prevents progression of heart failure. Circ 2002; 106:2491-2496.
56. Schäfer M, Frischkopf K, Taimor G et al. Hypertrophic effect of selective β₁-adrenoceptor stimulation on ventricular cardiomyocytes from adult rat. Am J Physiol Cell Physiol 2000; 279:C495-C503.
57. Schäfer M, Pönicke K, Heinroth-Hoffmann I et al. β-Adrenoceptor stimulation attenuates the hypertrophic effect of α-adrenoceptor stimulation in adult rat ventricular cardiomyocytes. J Am Coll Cardiol 2001; 37:300-307.
58. Wang Y, Huang S, Sah VP et al. Cardiac muscle cell hypertrophy and apoptosis induced by distinct members of the p38 mitogen-activated protein kinase family. J Biol Chem 1998; 273:2161-2168.

59. Crespo P, Cachero TG, Xu N et al. Dual effect of β-adrenergic receptors on mitogen-activated protein kinase. Evidence for a βγ-dependent activation of a $G\alpha_s$-cAMP-mediated inhibition. J Biol Chem 1995; 270:25259-25265.
60. Engelhardt S, Hein L, Keller U et al. Inhibition of Na^+-H^+ exchange prevents hypertrophy, fibrosis, and heart failure in β_1-adrenergic receptor transgenic mice. Circ Res 2002; 90:814-819.
61. Bisognano JD, Weinberg HD, Bohlmeyer TJ et al. Myocardial-directed overexpression of the human β_1-adrenergic receptor in transgenic mice. J Mol Cell Cardiol 2000; 32:817-830.
62. Iwata M, Yoshikawa T, Baba A et al. Autoimmunity against the second extracellular loop of β_1-adrenergic receptors induces β-adrenergic receptor desensitisation and myocardial hypertrophy in vivo. Circ Res 2001; 88:578-586.
63. Leineweber K, Heinroth-Hoffmann I, Pönicke K et al. Cardiac β-adrenoceptor desensitization due to increased β-adrenoceptor kinase activity in chronic uremia. J Am Soc Nephrol 2002; 13:117-124.
64. Mondry A, Bourgeois F, Carre F et al. Decrease in β_1-adrenergic and M2-muscarinic receptor mRNA levels and unchanged accumulation of mRNAs coding for $G\alpha_{i-2}$ and $G\alpha_s$ proteins in rat cardiac hypertrophy. J Mol Cell Cardiol 1995; 27:2287-2294.
65. Galinier M, Senard JM, Valet P et al. Cardiac β-adrenoceptors and adenylyl cyclase activity in human left ventricular hypertrophy due to pressure overload. Fundam Clin Pharmacol 1994; 8:90-99.
66. Böhm M, Kirchmayr R, Erdmann E. Myocardial $G_i\alpha$-protein levels in patients with hypertensive cardiac hypertrophy, ischemic heart disease and cardiogenic shock. Cardiovasc Res 1995; 30:611-618.
67. Asai K, Yang GP, Geng YJ et al. β-Adrenergic receptor blockade arrests myocyte damage and preserves cardiac function in the transgenic $Gs\alpha$ mouse. J Clin Invest 1999; 104:551-558.
68. Dubus I, Samuel JL, Marotte F et al. β-Adrenergic agonists stimulate the synthesis of noncontractile but not contractile proteins in cultured myocytes isolated from adult rat hearts. Circ Res 1990; 66:867-874.
69. Zhou XJ, Schlüter K-D, Piper HM. Hypertrophic responsiveness to β_2-adrenoceptor stimulation on adult ventricular cardiomyocytes. Mol Cell Biochem 1996; 163/164:211-216.
70. Simm A, Schlüter K-D, Diez C et al. Activation of p70 s6 kinase by β-adrenoceptor agonists on adult cardiomyocytes. J Mol Cell Cardiol 1998; 30:2059-2067.
71. Xiao RP. β-Adrenergic signalling in the heart: Dual coupling of the β2-adrenergic receptor to G_s and G_i protein. Sci STKE 2001; RE15.
72. Schlüter K-D, Frischkopf K, Flesch M et al. Central role for ornithine decarboxylase in β-adrenoceptor mediated hypertrophy. Cardiovasc Res 2000; 45:410-417.
73. Nakano M, Kanda T, Matsuzaki S et al. Effect of losartan, an AT1 selective angiotensin II receptor antagonist, on isoproterenol-induced cardiac ornithine decarboxylase activity. Res Commun Mol Pathol Pharmacol 1995; 88:21-30.
74. Du XJ, Autelitano DJ, Dilley RJ et al. β_2-Adrenergic receptor overexpression exacerebrates development of heart failure after aortic stenosis. Circulation 2000; 101:71-77.
75. Murphy RJ, Beliveau L, Gardiner PF et al. Nifedipine does not impede clenbuterol-stimulated muscle hypertrophy. Proc Soc Exp Biol Med 1999; 221:184-187.
76. Wong K, Boheler KR, Petrou M et al. Pharmacological modulation of pressureinduced cardiac hypertrophy: Changes in ventricular function, extracellular matrix, and gene expression. Circulation 1997; 96:2239-2246.
77. Schlüter K-D, Zhou XJ, Piper HM. Induction of hypertrophic responsiveness to isoproterenol by TGF-β in adult rat cardiomyocytes. Am J Physiol Cell Physiol 1995; 269:C1311-C1316.
78. Taimor G, Schlüter K-D, Frischkopf K et al. Autocrine regulation of TGFβ expression in adult cardiomyocytes. J Mol Cell Cardiol 1999; 31:2127-2136.
79. Wenzel S, Taimor G, Piper HM et al. Redox-sensitive intermediates mediate angiotensin II-induced p38 MAP kinase activation, AP-1 binding activity, and TGF-β expression in adult ventricular cardiomyocytes. FASEB J 2001; 15:2291-2293.

CHAPTER 8

Intracellular Signaling Through Protein Kinases in Cardiac Mechanotransduction

Peter H. Sugden*

Abstract

There is good evidence that stress-induced deformation of the cardiac myocyte can activate intracellular signaling pathways, though how this is brought about is still partly a mystery, some clues being provided by the present volume of reviews. The activation of these signaling pathways is thought to be instrumental in producing the changes in myocyte morphology, sarcomerogenesis, and gene expression that occur during hypertrophic growth. Reversible protein phosphorylation and dephosphosphorylation control a wide range biological responses, and hypertrophic growth is no exception. Specifically, there is evidence of a role for lipid-based signaling and protein kinase C in strain-induced signaling events. Activation of protein kinase C is probably instrumental in activating the extracellular signal-regulated kinase 1/2 cascade. However, other protein kinases are activated by strain: these included stress-activated protein kinases (c-Jun N-terminal kinases, p38-mitogen-activated protein kinases) and the Janus activated kinases. Apart from these, there is also evidence that the extracellular matrix, focal adhesion-based signaling and activation of the focal adhesion kinase may play a role in the response of myocytes to strain. The myocyte probably integrates the myriad messages from a variety of signaling pathways and this determines the overall biological response.

It is widely (though not unanimously) accepted that the adult cardiac myocyte is a terminally-differentiated cell, i.e., it is incapable of undergoing complete cycles of cell division. When, as a result of the heart being subjected to haemodynamic or other forms of overload, an increased mechanical load is placed upon the myocyte in vivo, it responds by increasing its myofibrillar complement and overall cell size. The most common experimental manoeuvre to elicit this response is to induce a pressure overload on the left ventricle by constricting the (thoracic) aorta. Overall, this hypertrophy of the contractile cells allows the heart to accommodate the increased loading. Because of its importance in pathophysiology, attempts have been made to simulate the situation ex vivo. Acute changes in the activation of signaling pathways can be studied in the perfused heart ex vivo, as can changes in the rate of protein synthesis and the very early effects on patterns of gene transcription, but this preparation does not survive for a sufficient length of time for there to be any change in myocyte size. Isolated myocytes from neonatal rat hearts can, when attached to deformable membranes coated with a suitable substrate (e.g., collagen, fibronectin), be 'stretched' either statically or phasically (usually at about 1 Hz) (see, for example, ref. 1-3). 'Stretch' is an imprecise term from a physical viewpoint and the term strain, which has a precise physical meaning, is preferred. Strain is the proportional deformation induced in a body by the application of a stress, which is a force. It is

*Correspondencing Author: Peter H. Sugden—NHLI Division (Cardiac Medicine), Faculty of Medicine, Imperial College of Science, Technology and Medicine, Flowers Building (Floor 4), Armstrong Road, London SW7 2AZ, U.K. Email: p.sugden@imperial.ac.uk

Cardiac Mechanotransduction, edited by Matti Weckström and Pasi Tavi.
©2007 Landes Bioscience and Springer Science+Business Media.

assumed that the strained myocyte ex vivo simulates haemodynamic overloading in vivo, but this is not entirely justified. For example, rates of stretching of rat myocytes ex vivo are always less than the 5-6 Hz in vivo, since it is not possible to achieve such rates ex vivo. Furthermore, increased force of contraction in the whole heart is mediated by an increase in intracellular Ca^{2+} (Ca^{2+}_i) transient and/or by an increase in the sensitivity of the myofibrillar ATPase to Ca^{2+}_i. For technical reasons, it is still not clear whether strain increases Ca^{2+}_i, or the Ca^{2+}_i transient. However, the strained myocyte is still perhaps the best controlled experimental system and the myocytes survive for a sufficient period to allow changes in gene expression (which parallel those in vivo) to be detected.[1-4] Increased contractile activity can also be induced either by increasing cell density to cause increased spontaneous contraction,[5] or induced by electrical stimulation at about 3 Hz,[6] though again it is not clear to which in vivo processes these are analogous. Furthermore, it is still not clear how the myocyte detects the increases in mechanical loading or contractile activity, and how the 'mechanical sensors' couple to the signaling pathways responsible for sarcomerogenesis and cell growth. In lower organisms, environmental signals (osmolarity, chemotactic signals, etc.) are frequently detected by the histidine kinase 'two component' systems, but establishing that these exist in mammalian systems has proved difficult.[7,8] In higher organisms, the process of reversible phosphorylation and dephosphorylation of Ser-, Thr- and/or Tyr- residues in proteins by protein kinases and phosphatases is central to the regulation of numerous biological responses, and cell growth is no exception. In this article, I will summarise the state of current knowledge relating to the roles of protein phosphorylation and dephosphorylation in cardiac myocyte hypertrophy, and how these might be related to mechanotransduction.

'Extrinsic' and 'Intrinsic' Mechanotransduction

A number of mechanisms have been put forward to account for mechanotransductional growth and these have been extensively reviewed.[4] Because mechanical strain induces hypertrophy in isolated cultured myocytes, it is generally considered that the mechanical sensors which detect increased mechanical loads are inherent to individual myocytes and there is no obligatory requirement for the participation of signals from other cell types. Hypertrophic signals may emanate from outside or inside the cell. 'Extrinsic' signaling includes signaling dependent on the release of factors produced by the myocyte that act locally (autocrine/paracrine mechanisms),[4] signaling based on the interaction of the myocyte with the extracellular matrix (ECM) through integrins (or other cell surface molecules) and the cytoskeleton,[4,9,10] and signaling involving the activation of mechanosensitive ('stretch-activated') ion channels.[4,11] Because of its role in excitation-contraction coupling, Ca^{2+}-dependent signaling remains a candidate intermediary for hypertrophic growth, and, though also regulated extrinsically, could be classed as an 'intrinsic' pathway.

Endpoints of Myocyte and Myocardial Hypertrophy

Because myocardial growth allows the heart to accommodate increased mechanical loading, it is implicit that this is at least initially an adaptive and beneficial response. Myocyte hypertrophy is much easier to define and study in isolated cells than in the intact heart. Hypertrophy of isolated myocytes involves increases in cell size, protein content, sarcomerogenesis and cell-cell contacts. There are also changes in the patterns and rates of gene transcription, particularly with respect to the reexpression of 'foetal' genes such as atrial natriuretic peptide (ANF) and B-type natriuretic peptide (BNP), though instances of a lack of connection between these changes and the morphological changes do exist. Changes in transcription and morphology have been demonstrated in strained myocytes ex vivo.[1-4] Electrophysiologically, the action potential and the Ca^{2+} transient become prolonged in the hypertrophied myocyte, probably because of changes in the relative abundances of proteins regulating the movements of ions. Many of these changes are also seen in vivo. However, in the in vivo situation, the term 'hypertrophy' tends to be applied at the whole organ level rather than at the cellular level and can additionally include maladaptive cardiac enlargement, which leads to confusion. Probably

Table 1. Protein kinases and phosphatases that may participate in cardiac myocyte hypertrophy

1. Protein kinase C isoforms
2. Mitogen-activated protein kinases (MAPKs)
    (a) Extracellular signal-regulated kinases 1/2
    (b) Stress activated protein kinases
        (i) c-Jun N-terminal kinases
        (ii) p38 mitogen-activated kinases
        (iii) 'big' MAPK1
3. Phosphoinositide 3-OH kinase-dependent signalling
    (a) Akt (protein kinase B)
    (b) Glycogen synthase kinase 3
4. Ca^{2+}-dependent signalling
    (a) Ca^{2+}/calmodulin-dependent kinases
    (b) Myosin light chain kinase
    (c) Calcineurin (protein phosphatase 2A)
5. Rho-dependent kinase pathways
6. Janus-activated kinases

no single endpoint is satisfactory, and the preferred approach should be to characterize the phenotype as fully as possible. In some cases, inappropriate end points of 'hypertrophy' have been chosen (e.g., activation of signaling proteins that may be associated with the induction of hypertrophy)[12] and this is a practice to be discouraged.

Signaling in Mechanotransduction and Hypertrophy: Protein Kinase C

Many signaling pathways have been implicated in the hypertrophic response of the cardiac myocyte (Table 1) and it is likely that there is considerable redundancy in signaling and considerable interaction between the various pathways. The first protein kinases to be linked with cardiac myocyte hypertrophy were the protein kinase C (PKC) family.[13,14] All PKCs can be structurally subdivided into an N-terminal regulatory domain (containing the autoinhibitory 'pseudosubstrate' site, and the cofactor/activator binding sites), a 'hinge' region that is susceptible to proteolysis, and a C-terminal catalytic region. PKCs are dependent on phospholipids (particularly phosphatidylserine) for activity and can be divided in three subfamilies depending on their further requirements for cofactors and activators. The 'classical' or 'conventional' cPKCs require diacylglycerol (DAG) and Ca^{2+} in addition to phosphatidylserine, and the 'novel' nPKCs require DAG though probably not Ca^{2+}. The activities if the 'atypical' aPKCs appear to be independent of both DAG and Ca^{2+}, and their regulation remains poorly-understood. The nPKC subfamily is sometimes further subdivided depending on the presence or absence of a domain with homology to the Ca^{2+}-binding (C2) domain of the cPKCs. Transfection of cardiac myocytes with plasmids encoding primarily the catalytic regions of cPKCα and cPKCβ induced transcriptional changes typical of hypertrophy,[15,16] and it was shown later that small activating mutations in PKC holoenzymes could also elicit similar responses.[17]

Activation of cPKCs and nPKCs involves hydrolysis of the membrane phospholipid phosphatidylinositol 4',5'-bisphosphate [PtdIns(4,5)P$_2$] (Fig. 1). Under the influence of the binding of a variety of agonists to their transmembrane receptors [especially those of the heterotrimeric Gq protein-coupled receptor class, e.g., the endothelin (ET) type A receptor], phospholipase Cβ is activated and PtdIns(4,5)P$_2$ is hydrolyzed to hydrophobic DAG, which remains in the plane of the membrane, and the soluble second messenger inositol 1,4,5-trisphosphate. This hydrolysis can also be effected by phospholipase Cγ, which is normally regulated through a different class of transmembrane receptors, the receptor protein Tyr-kinases. As the plasma membrane content of DAG increases, DAG-sensitive PKCs

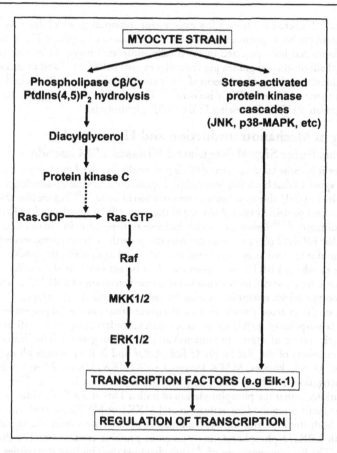

Figure 1. Signaling through the mitogen-activated and stress-activated protein kinase cascades. Mechanical strain, in a way that is not understood, is translated to stimulate phosphatidylinositol (4,5)bisphosphate [PtdIns(4,5)P$_2$] hydrolysis and to activate protein kinase C. In a way that is not understood, this leads the activation of the small G protein Ras and the extracellular signal-regulated kinase 1/2 (ERK1/2) cascade. ERK1/2 then phosphorylate transcription factors to change their transactivating activities and thus alter the rates and patterns of transcription. In addition, strain activates the c-Jun N-terminal kinases and the p38 mitogen-activated protein kinases (though this process is poorly-understood), and their activation may alter transcriptional rates and profiles.

translocate from the cytoplasm and associate with the membrane fraction and this is thought to be indicative of their activation. Tumor-promoting phorbol esters such as phorbol 12-myristate 13-acetate (PMA) mimic DAG, although they produce an unphysiological long-lasting translocation of cPKCs and nPKCs to the membrane compared with 'physiological' agonists. Over the longer term, they deplete also the cell of the cPKCs and nPKCs, and this is usefully experimentally from the point of view of inhibiting cPKC/nPKC-dependent responses. In spite of the differences from 'physiological' agonists, phorbol esters have been extensively and usefully utilized in implicating cPKCs and nPKCs in biological processes.

Increased hydrolysis of PtdIns(4,5)P$_2$ has been detected in mechanically-strained myocytes,[18,19] and PKC has been implicated in the strain-induced expression of the *c-fos* gene.[18] *c-fos* encodes the c-Fos transcription factor which may be involved in the induction of the transcriptional changes in hypertrophy. Increased PtdIns(4,5)P$_2$ hydrolysis has been detected in perfused rodent hearts 'stretched' by increasing the aortic pressure or by balloon dilatation of the ventricle.[20,21] There have been a few experiments reported in which increased contractile activity or strain has been shown to increase the membrane-association of PKCs. Translocation of PKC isoforms has been detected in electrically-stimulated myocytes ex vivo (nPKCδ and nPKCε),[6] in balloon-dilated guinea pig ventricles ex vivo (nPKCε),[21] and in guinea pig hearts haemodynamically overloaded by 4 weeks of aortic constriction in vivo (cPKCα and, surprisingly, the nervous tissue restricted cPKCγ isoform).[22] These experiments suggest that mechanical strain can activate PKC via increasing PtdIns(4,5)P$_2$ hydrolysis.

Signaling in Mechanotransduction and Hypertrophy: The Extracellular Signal-Regulated Kinases 1/2 Cascade

Soon after it became apparent that PKC is potentially involved in myocyte hypertrophy, it also became apparent that bona fide hypertrophic agonists such as the α$_1$-adrenergic agonists[23,24] and ET-1[25] are not only able to stimulate translocation of nPKCs,[26] but are also able to activate mitogen-activated protein kinases (MAPKs) of the extracellular signal-regulated kinase 1 and 2 (ERK1/2) subfamily.[27,28] Given the established role of these protein kinases in cell growth, a hypothesis that ERK1/2 plays a role in the adaptive growth of the myocyte was advanced.[27] A large amount of evidence has been presented both for and against this hypothesis,[29] but recent experiments in which ERK1/2 were engineered to be activated in the cardiac myocytes of transgenic mice have convincingly demonstrated that activation of ERK1/2 in vivo induces a cardiac phenotype which resembles an adapted pressure overload hypertrophy.[30]

ERK1/2 are the 'effector' protein kinases of a linear three-membered protein kinase cascade (Fig. 1).[31] The purposes of such linear protein kinase cascades are signal amplification (because of the catalytic nature of their components) and signal integration. The 'initiator' protein kinases are members of the Raf family (c-Raf, A-Raf and B-Raf), which phosphorylate the 'intermediate' protein kinases, MAPK kinases 1 or 2 (MKK1/2), on 2 Ser- residues, thereby producing activation. MKK1/2 then phosphorylate and activate ERK1/2. These are unusual phosphorylations in that the phosphorylation of both a Thr- and a Tyr- residue in an ERK1/2 Thr-Glu-Tyr motif is required for activation and MKK1 or MKK2 are both capable of phosphorylating both the Thr- and the Tyr- residues. After their activation, the signaling cascade diverges with ERK1/2 phosphorylating numerous proteins contain Ser/Thr-Pro (or better Pro-Xaa-Ser/Thr-Pro) consensus motifs,[32] thus changing their biological activities. Established substrates for ERK1/2 include transcription factors (e.g., Elk-1),[33] protein kinases [e.g., 90-kDa ribosomal protein S6 kinases (RSKs)],[34] and other signaling proteins (e.g., cytoplasmic phospholipase A$_2$).[35] In myocytes, ERK1/2 are rapidly but transiently activated by agonists such as ET-1 or PMA,[28] and activated ERK1/2 rapidly appear in the nucleus.[36] Inactivation involves their dephosphorylation, but the phosphatases responsible are not well-characterised and may include protein Ser-/Thr- phosphatases (e.g., PP1, PP2A), protein Tyr phosphatases, or more specialised MAPK dual specificity phosphatases which may be either constitutively- or inducibly-expressed.

Both ET-1 and PMA are powerful activators of the ERK1/2 cascade in the cardiac myocyte.[27,28,37] Since both induce PKC translocation,[26,38] a connection between activation of PKC and the ERK1/2 cascade seems likely. c-Raf is normally cytoplasmic, but migrates to the plasma membrane when the membrane-bound small (21 kDa) G proteins of the Ras family[39,40] are activated. Ras is often thought of as being a 'molecular switch' (Fig. 2). In its 'off' state, it is ligated to GDP, activation involves exchange of GDP for GTP, and this exchange reaction is stimulated by guanine nucleotide exchange factors (GEFs) such as Sos. The Ras.GTP switch is 'turned off' by the innate GTPase activity of Ras, which is low under basal conditions but is significantly stimulated by GTPase activating proteins (GAPs). Thus Ras is potentially activated by stimulation of GEF activity or inhibition of GAP activity, but, in order to induce the rapid GTP loading of Ras that is seen in the cardiac myocytes in response to ET-1 or PMA,[36] it is kinetically more likely that a GEF is activated than a GAP inhibited. c-Raf does not bind to Ras.GDP but interacts strongly with Ras.GTP through its N-terminal Ras

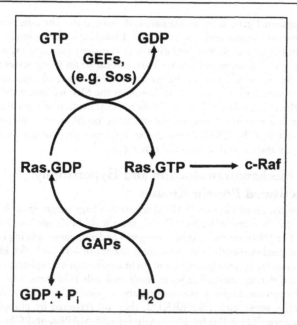

Figure 2. The Ras guanine nucleotide exchange cycles. In its 'off' state, the small G protein Ras is in its GDP-ligated form. Activation of guanine nucleotide exchange factors (GEFs) stimulates exchange of GTP for GDP, to produce the 'on' species, which then participates in the activation of c-Raf. The innate GTPase activity of Ras, an activity that is greatly stimulated by GTPase-activating protein (GAPs) returns Ras.GTP to the 'off' state.

interaction domain, and thus migrates to the membrane.[39,40] Although binding of c-Raf to Ras.GTP is probably not sufficient to activate the kinase fully, other events at the membrane (possibly involving phosphorylation of c-Raf, particularly Tyr-phosphorylation) bring about full activation.[41,42]

There has been surprising little work on the effects of haemodynamic overload on activation of the ERK1/2 cascade in vivo. ERK1/2 are acutely (10-30 min) stimulated following aortic constriction in rodents, but it is difficult to gauge the extent of activation relative to the maximum achievable.[43,44] Some workers have obtained similar results in rodents over more prolonged periods (from 7 h to 4 weeks, after which the hearts begin to fail),[45,46] though this result has not been confirmed by others.[44] There has been relatively little work reported for the isolated perfused heart, though Takeishi et al have shown that balloon dilation of the left ventricle acutely activates ERK1/2 and one group of their effectors, the RSKs.[46] Very similar results have been obtained with mechanically-strained myocytes ex vivo, where it is clear that many of the components of the ERK1/2 signaling pathway [PtdIns(4,5)P$_2$ hydrolysis, PKC, Ras, ERK1/2] and downstream effectors of the ERK1/2 cascade (RSKs, Elk-1) become activated.[19,47]

The changes in the activity of the ERK1/2 cascade potentially have a direct impact on gene transcription. For example, 'ternary complex' transcriptional signaling becomes activated.[33] The ternary complex consists of Elk-1, the serum response factor (SRF) and a oligonucleotide sequence containing regulatory consensus recognition sequences to which Elk-1 and SRF bind. Such regulatory regions usually lie 5' to the transcriptional initiation site of genes and can alter the rate of transcription of susceptible genes. The oligonucleotide sequence to which SRF binds is known as the serum response element (SRE). There are multiple sites in the C-terminal region of Elk-1 that are phosphorylated by ERK1/2 (and possibly other protein kinases, such as the stress-activated protein kinases, see below).[48] Some or all of these phosphorylations increase the DNA binding- and transactivating-activity of Elk-1.[33] The epitome of ternary complex factor-regulated gene is the immediate early gene c-*fos*. The c-*fos* promoter contains neighbouring consensus sequences for SRF and Elk-1 binding in the region lying -300 to -330 bp upstream from the transcriptional initiation site. Transient increased expression of c-*fos* is one of the early events following induction of mechanical strain in myocytes ex vivo,[1,2] or in the overloaded heart in vivo,[49] and, as already mentioned, increased expression of c-*fos* may participate in the hypertrophic response. Furthermore, the SRE represents the 'pressure sensitive' element in the c-*fos* promoter in vivo.[50] Given the activation of the ERK1/2 cascade by mechanical strain, it is easy to see how mechanical strain could induce c-*fos* expression. In the same way, it is likely that the ERK1/2 cascade can influence the biological activities of other transcription factors and the transcription of other genes.

Signaling in Mechanotransduction and Hypertrophy: The Stress-Activated Protein Kinases

The stress-activated protein kinases (SAPKs) represent a large subgroup of the MAPK family, and include the c-Jun N-terminal kinases (JNKs), the p38-MAPKs and 'big' MAPK 1 (BMK1 or ERK5).[31,51,52] The JNKs are encoded by three genes and alternative splicing of the transcripts produces proteins predominantly with molecular masses of ~46 and ~54 kDa.[53] Likewise, p38-MAPKs are encoded by four genes, two of them are alternatively spliced, and the isoforms display somewhat differing susceptibilities to small molecule inhibitors. The SAPKs are activated primarily by cytotoxic cellular stresses (metabolic poisons, protein synthesis inhibitors, osmotic or oxidative stress, etc). Like ERK1/2, they are activated by phosphorylation of a Thr-Xaa-Tyr sequence (Xaa is Pro for JNKs, Gly for p38-MAPKs, and Glu for ERK5) and they also phosphorylate Ser/Thr-Pro sequences in proteins (including transcription factors and other protein kinases). They are also the 'effector kinases' of (probably) three-membered cascades, though the structures of these cascades are less well characterised than for the ERK1/2 cascade. Like the ERK1/2 cascade, all of the SAPK cascades have been implicated in myocardial hypertrophy,[54,55] though the evidence is less convincing than for ERK1/2 in my view. Certainly, there is no indication that activation of SAPKs in the hearts of transgenic mice by activating the MKK stage produces anything other than a morbid phenotype.[30,55]

Aortic constriction in vivo or increased perfusion pressure in isolated hearts both acutely and chronically increase activities of JNKs and p38-MAPKs (though one group failed to detect an increase in JNK activity in vivo or following balloon dilation of the left ventricles of isolated hearts),[46] and the proportional increases may be significantly greater than for ERK1/2.[43-46,56] Caution needs to be exercised in the interpretation of these results. Ischaemia followed by reperfusion is a powerful activator of JNKs and p38-MAPKs in the heart,[57] and it is possible that the intermittent subendocardial ischaemia associated with increased aortic pressure leads to activation of the JNKs and p38-MAPKs. However, mechanical strain activates JNKs and p38-MAPKs in isolated myocytes,[58,59] a situation where delivery of O_2, metabolic fuels, etc. should not be limiting. Electrical pacing of isolated neonatal myocytes also increases phosphorylation of 54 kDa JNKs, as well as an unusual 49 kDa JNK species, but does not increase phosphorylation of the 46 kDa species.[6] Increasing methodological sophistication has allowed strain to be imposed at different stages in the cardiac cycle.[59] The rationale underlying these experiments is that the strain induced by haemodynamic pressure overload of the heart in vivo occurs primarily at systole, whereas strain induced by volume overload occurs primarily during diastole. The two forms of strain produce different morphological responses in the heart in vivo (ventricular wall thickening in the absence of chamber enlargement in the case of pressure overload, increased chamber diameter with a proportional increase in ventricular wall thickness in the case of volume overload).[60] In the case of pressure overload, sarcomeres are laid down in primarily a parallel fashion whereas, in volume overload, there is a significant in series component.[60] Imposing strain on myocytes at either the contraction (systolic) or relaxation (diastolic) stage activates JNKs to the same extent, and similar results were obtained for p38-MAPKs.[59] In contrast, the ERK1/2 cascade and expression of BNP are more powerfully activated by systolic strain than diastolic strain.[59] Though it is a little difficult to extrapolate these results to the in vivo situation, it is possible to speculate that the different patterns of MAPK/SAPK activation could lead to the different patterns of sarcomerogenesis seen in pressure or volume- overload. In contrast to the relatively extensive studies of JNKs and p38-MAPKs, very little work has been carried out on BMK1, though Takeishi et al showed that it becomes phosphorylated (activated) following aortic constriction in vivo.[46]

'Focal Adhesion'-Associated Signaling: Basic Information

Myocytes interact with the ECM in a number of ways. Generally, plating of cultured cells on a suitable ECM (e.g., fibronectin) causes the engagement of their integrins with their extracellular ligands and/or integrin clustering, stimulating the formation of 'focal adhesions' or 'focal complexes'.[10,61,62] In myocytes, these focal adhesions may be stabilised by contractile activity.[63] There is still discussion of whether focal adhesions exist as such in vivo or whether they are simply induced under the rather artificial conditions of cell culture. However, overall, integrin engagement/clustering and focal adhesion formation lead to the activation of intracellular signaling pathways, including the ERK1/2 cascade.[64] Integrins are ubiquitous heterodimeric ($\alpha\beta$) transmembrane glycoproteins, which in addition to being involved in cell adhesion, possess bidirectional signaling functions and transmit extracellular signals to the intracellular compartment and vice versa.[9,10] They are selective in terms of their interaction with the ECM. The $\alpha_5\beta_1$ integrin (which is expressed in cardiac myocytes)[9] is a major receptor for fibronectin, recognizing an -Arg-Gly-Asp- (RGD) containing sequence,[65,66] and oligopeptides containing the RGD sequence interfere with these interactions.

Focal adhesions themselves consist of numerous proteins in addition to integrins including structural cytoskeletal proteins and a wide range of signaling proteins. The latter group includes protein Tyr- kinases such as Src family kinases (which are activated following aortic constriction in vivo)[46] and the c-Src regulatory kinase, CSK,[67,68] focal adhesion kinase (FAK),[69-71] and the FAK-related Pro-rich tyrosine kinase 2 (PYK2)[72] (also known as RAFTK, CAKβ, or CADTK), a Ca^{2+}-dependent protein Tyr- kinase. Other focal adhesion-associated signaling proteins include protein Ser-/Thr- kinases [e.g., PKC (see above), p21-activated kinase (PAK)],[73,74] small G protein modulators, protein and lipid phosphatases (e.g., protein Tyr- phosphatases and PTEN),[75] and docking and adapter proteins. Many of these are involved in the regulation, transmission, or termination of signaling events. Thus, the docking protein p130Cas binds directly to Pro-rich regions in FAK and PYK2, and it becomes Tyr-phosphorylated on activation of FAK or PYK2.[76] Other signaling proteins may bind to phospho-p130Cas through their SH2 domains (which selectively recognise phospho-Tyr-containing sequences in proteins). Adapter/scaffolding proteins such as paxillin, itself a protein Tyr- kinase substrate, are also present.[77,78] The association of docking and adaptor molecules with FAK and PYK2 and their phosphorylation provide a focus for the binding of further signaling molecules. Thus, in one model of focal adhesion-based signaling, the formation of focal adhesions induces the autophosphorylation of Tyr^{397} in FAK, a phosphorylation that may require an intact actin cytoskeleton and activation of the small G protein, Rho. FAK(phospho-Tyr^{397}) then recruits membrane-bound c-Src by binding of the c-Src SH2 domain to the FAK(phospho-Tyr^{397}) sequence.[67,68] In catalytically-inactive c-Src, this SH2 domain is ligated intramolecularly to c-Src(phospho-Tyr^{527}), phosphorylation of which is catalyzed by c-Src kinase, CSK, but competition from p125-FAK(phospho-Tyr^{397}) disrupts this interaction. This not only 'relaxes' the c-Src structure leading to dephosphorylation of c-Src(phospho-Tyr^{527}), but also stimulates autophosphphosphorylation of c-Src(Tyr^{416}) and activation of c-Src. c-Src then phosphorylates FAK(Tyr^{925}), thus creating a docking site for the docking protein Grb2 of the Grb2-Sos complex. In a manner analogous to receptor protein Tyr kinase signaling,[79] these events place the Sos GEF in the proximity of membrane-bound Ras.GDP. The enhancement of GDP/GTP exchange by Sos activates Ras, leading to activation of the ERK1/2 cascade.

Signaling in Mechanotransduction and Hypertrophy: Focal Adhesion-Based Signaling

The situation with respect to the participation of ECM-based signaling in the hypertrophic response is complex. Mechanically-strained myocytes release autocrine and paracrine factors such as the vasoactive peptides ET and angiotensin II (ANGII),[12,80,81] which then elicit signaling events in the myocyte that has secreted them (autocrine signaling) or in other myocytes (paracrine signaling). These peptides (particularly ET-1) activate the MAPK/SAPK signaling pathways that were described earlier,[82] and are established hypertrophic agonists in their own right.[83] Using BNP expression as a criterion of the hypertrophic response, experiments with cyclically-strained myocytes adhering to a fibronectin matrix showed that, although there is a significant ANGII type 1 (AT1) receptor- and ET type A receptor-dependent effect of strain on BNP expression, an AT1/ET$_A$ receptor-independent component remains. Thus, strain-induced BNP expression is only inhibited by ~60% using a combination of an AT-1 receptor antagonist (losartan) and an ET$_A$ receptor antagonist (BQ123), presumably both at maximally-effective concentrations.[81] In contrast, RGD peptides or soluble fibronectin causing a complete inhibition of strain-induced BNP expression, and certain anti-integrin antibodies were capable of partially inhibiting the response alone or, with increased efficacy, in combination.[84,85] Both the strain dependent- and the strain-independent-components of BNP expression involve ERK1/2 and the p38-MAPKs.[84,86] Activation of p38-MAPKs appears to be at least partly dependent on focal adhesion-associated proteins (e.g., PTEN, FAK, CSK).[87] The conclusion is that the ECM participates in the strain-dependent release of autocrine/paracrine factors as well as directing purely strain-dependent signaling. However, the experiments with RGD peptides, exogenous fibronectin or anti-integrin antibodies are not simple to interpret. It is presumably implicit that strain-dependent growth must involve ECM attachment whatever the mechanism because ECM attachment is the intermediary between the myocyte and the deformable membrane substrate. If this is wholly or partially disrupted, then it could interfere with the transmission of the deforming force, but that does not necessarily mean that there is active participation of integrin-based signaling in strain induced growth. Integrin/focal complex-based signaling has also been implicated in the pressure overloaded ventricle in vivo,[88,89] though, in this situation, the cellular heterogeneity of the heart is a complicating factor and it is not clear whether this signaling pathway is sufficient on its own to induce the response.

Signaling in Mechanotransduction and Hypertrophy: Other Protein Kinases (and Phosphatases)

Other protein kinases potentially involved in mechanotransduction and myocardial hypertrophy can be divided into 3 classes, those which are associated with activation of the phosphoinositide 3'-OH kinase signaling pathway [Akt,[90] glycogen synthase kinase 3 (GSK3)],[91,92] those which are activated by increased Ca^{2+}$_i$ and the Ca^{2+}$_i$-sensing protein, calmodulin (CaM) [Ca^{2+}/CaM kinase,[93] myosin light chain kinase],[94] and the Janus activated kinase/signal transducer and activator of transcription (JAK/STAT) pathway.[95,96] Though they will not be discussed here (because their role in mechanotransduction has not yet been studied), kinases that are dependent on the active form of the small G protein RhoA[97] may be important in the hypertrophic response, possibly in part through their action on the SRE.[98,99]

As mentioned earlier, Ca^{2+}$_i$ is a viable candidate as a 'second messenger' that couples increased myocardial contractility to hypertrophy. Blockade of the L-type Ca^{2+} channel (which is partly responsible for the Ca^{2+}$_i$ transient during contraction) reduces myocardial hypertrophy that develops with age in the spontaneously-hypertensive rat,[100] though it is not entirely clear that this is related to anything other than the antihypertensive properties of the L-type Ca^{2+} channel antagonists. With respect to mechanotransduction and hypertrophy (although it catalyzes dephosphorylation of proteins rather than their phosphorylation), there has been more work reported on the Ca^{2+}$_i$/CaM-activated protein phosphatase, calcineurin (CaN, or protein phosphatase 2B)[101,102] than on Ca^{2+}/CaM-dependent kinases (Ca^{2+}/CaM kinase and myosin light chain kinase). The participation of these kinases in mechanotransduction remains largely hypothetical. The situation for CaN is different. CaN is inhibited by the immunosuppressants cyclosporin or FK506. Though the verdict is not unanimous, several groups have independently established that these compounds prevent myocardial hypertrophy induced by aortic constriction.[102] In transgenic mice engineered to overexpress an endogenous protein inhibitor of CaN (Cain/Cabin) in a cardiomyocyte-specific manner, or in rats whose hearts overexpressed Cain/Cabin following adenoviral transfer, myocardial hypertrophy is reduced following aortic constriction.[103] The interpretation of these experiments is that an increased Ca^{2+}_i transient and/or an increase in diastolic Ca^{2+}_i activate CaM/CaN signaling, leading to hypertrophy. To my knowledge, no analogous experiments have been performed in mechanically-strained myocytes.

Unlike most protein kinases, GSK3 is a tonically-active kinase which is inhibited by phosphorylation mediated by Akt, the protein kinase kinase whose activity is regulated by the phosphoinositide 3-OH kinase pathway.[104] Inhibition of GSK3 promotes cardiac myocyte hypertrophy.[91] In transgenic mice engineered to express (in a cardiomyocyte-specific manner) a mutant species of GSK3β that is not susceptible to Akt-mediated inhibition, the development of myocardial hypertrophy following aortic constriction is retarded,[105] and a Fas death receptor-mediated inhibition of GSK3β may be involved in the response of the heart to pressure overload.[106] To my knowledge, no analogous experiments have been performed in mechanically-strained myocytes.

JAKs are protein Tyr- kinases that phosphorylate STAT transcription factors, which then dimerise and enter the nucleus to regulate the expression of sensitive genes.[95,96] The JAK/STAT pathway is primarily associated with cytokine signaling.[96] In the myocyte, its activation been associated cardiotrophin-induced hypertrophy (which utilises a gp130/cytokine-like signaling system),[107] and also interconnects with signaling by ANGII and the ERK1/2 cascade.[95,96] As recently reviewed by Booz et al[95] there is evidence of activation of the JAK/STAT pathway by acute pressure overload in vivo,[108] and in mechanical-strained myocytes ex vivo.[44,109]

General Conclusions

It is accepted that protein phosphorylation and dephosphorylation are involved in the development of cardiac myocyte hypertrophy, and, in my view, the clearest evidence of involvement is of the ERK1/2 cascade and CaN. What is still not clear is how the myocyte signaling pathways are activated by strain. Release of small molecules (prostaglandins, reactive oxygen species) or peptide mediators may activate the signaling pathways in an autocrine or paracrine manner, but acceptance of that merely changes the question to one of how do the systems that control the synthesis and release of these agonists detect strain? A good case can be made for focal adhesion signaling, though an understanding of the multiple protein-protein interactions is still somewhat distant. Ca^{2+}_i-dependent signaling remains a strong candidate but it is unclear whether CaM, the Ca^{2+}_i-sensor, becomes Ca^{2+}-ligated and activated during the contractile cycle. Mechanosensitive ion channels also remain candidates, though progress in this direction has been relatively limited over the last few years. The discovery of novel peptide blockers of these channels may facilitate progress.[110] Thus, although it is confusing for us mere mortals, the myocyte is probably cleverer, and it understands the myriad messages from a variety of signaling pathways and integrates them into an overall biological response.

References

1. Komuro I, Kaida T, Shibazaki Y et al. Stretching cardiac myocytes stimulates protooncogene expression. J Biol Chem 1990; 265:3595-3598.
2. Sadoshima J, Jahn L, Takahashi T et al. Molecular characterization of stretch-induced adaptation of cultured cardiac cells. An in vitro model of load-induced cardiac hypertrophy. J Biol Chem 1992; 267:10551-10560.
3. Liang F, Wu J, Garami M et al. Mechanical strain increases expression of the brain natriuretic peptide gene in rat cardiac myocytes. J Biol Chem 1997; 272:28050-28056.

4. Sadoshima J, Izumo S. The cellular and molecular response of cardiac myocytes to mechanical stress. Annu Rev Physiol 1997; 59:551-571.
5. Eble DM, Cadre BM, Qi M et al. Contractile activity modulates atrial natriuretic factor gene expression in neonatal rat ventricular myocytes. J Mol Cell Cardiol 1998; 30:55-60.
6. Strait JB, Samarel AM. Isoenzyme-specific protein kinase C and c-Jun N-terminal kinase activation by electrically stimulated contraction of neonatal rat ventricular myocytes. J Mol Cell Cardiol 2000; 32:1553-1566.
7. Klumpp S, Krieglstein J. Phosphorylationa and dephosphorylation of histidine residues in proteins. Eur J Biochem 2002; 269:1067-1071.
8. Steeg PS, Palmieri D, Ouatas T et al. Histidine kinases and histidine phosphorylated proteins in mammalian cell biology, signal transduction and cancer. Cancer Lett 2003; 190:1-12.
9. Ross RS, Borg TK. Integrins and the myocardium. Circ Res 2001; 88:1112-1119.
10 Juliano RL. Signal transduction by cell adhesion receptors and the cytoskeleton: Functions of integrins, cadherins, selectins and immunoglobulin-superfamily members. Annu Rev Pharmacol Toxicol 2002; 42:283-323.
11. Hu H, Sachs F. Stretch-activated ion channels in the heart. J Mol Cell Cardiol 1997; 29:1511-1523.
12. Yamazaki T, Komuro I, Kudoh S et al. Endothelin-1 is involved in mechanical stress-induced cardiomyocyte hypertrophy. J Biol Chem 1996; 271:3221-3228.
13. Nishizuka Y. Protein kinase C and lipid signaling for sustained cellular responses. FASEB J 1995; 9:484-496.
14. Newton AC. Protein kinase C: Structural and spatial regulation by phosphorylation, cofactors, and macromolecular interactions. Chem Rev 2001; 101:2353-2364.
15. Kariya K, Karns LR, Simpson PC. Expression of a constitutively activated mutant of the β-isozyme of protein kinase C in cardiac myocytes stimulates the promoter of the β-myosin heavy chain isogene. J Biol Chem 1991; 266:10023-10026.
16. Shubeita HE, Martinson EA, van Bilsen M et al. Transcriptional activation of the cardiac myosin light chain 2 and atrial natriuretic factor genes by protein kinase C in neonatal rat ventricular myocytes. Proc Natl Acad Sci USA 1992; 89:1305-1309.
17. Decock JBJ, Gillespie-Brown J, Parker PJ et al. Classical, novel and atypical isoforms of PKC stimulate ANF- and TRE/AP-1-regulated-promoter activity in ventricular cardiomyocytes. FEBS Lett 1994; 356:275-278.
18. Komuro I, Katoh Y, Kaida T et al. Mechanical loading stimulates cell hypertrophy and specific gene expression in cultured rat cardiac myocytes. Possible role of protein kinase C activation. J Biol Chem 1991; 266:1265-1268.
19. Sadoshima J, Izumo S. Mechanical stretch rapidly activates multiple signal transduction pathways in cardiac myocytes: Potential involvement of an autocrine/paracrine mechanism. EMBO J 1993; 12:1681-1692.
20. Bogoyevitch MA, Fuller SJ, Sugden PH. Cyclic AMP and protein synthesis in isolated adult rat heart preparations. Am J Physiol 1993; 265:C1247-C1257.
21. Paul K, Ball NA, Dorn II GW et al. Left ventricular stretch stimulates angiotensin II-mediated phosphatidylinositol hydrolysis and protein kinase C ε isoform translocation in adult guinea pig hearts. Circ Res 1997; 81:643-650.
22. Jalili T, Takeishi Y, Song G et al. PKC translocation without changes in $G\alpha_q$ and PLC-β protein abundance in cardiac hypertrophy and failure. Am J Physiol 1999; 277:H2298-H2304.
23. Simpson P. Norepinephrine-stimulated hypertrophy of cultured rat myocardial cells is an α_1 adrenergic response. J Clin Invest 1983; 72:732-738.
24. Lee HR, Henderson SA, Reynolds R et al. α_1-adrenergic stimulation of cardiac gene transcription in neonatal rat myocardial cells. Effects on myosin light chain-2 gene expression. J Biol Chem 1988; 263:7352-7358.
25. Shubeita HE, McDonough PM, Harris AN et al. Endothelin induction of inositol phospholipid hydrolysis, sarcomere assembly, and cardiac gene expression in ventricular myocytes. A paracrine mechanism for myocardial cell hypertrophy. J Biol Chem 1990; 265:20555-20562.
26. Clerk A, Bogoyevitch MA, Andersson MB et al. Differential activation of protein kinase C isoforms by endothelin-1 and phenylephrine, and subsequent stimulation of p42 and p44 mitogen-activated protein kinases in ventricular myocytes cultured from neonatal rat hearts. J Biol Chem 1994; 269:32848-32857.
27. Bogoyevitch MA, Glennon PE, Sugden PH. Endothelin-1, phorbol esters and phenylephrine stimulate MAP kinase activities in ventricular cardiomyocytes. FEBS Lett 1993; 317:271-275.

28. Bogoyevitch MA, Glennon PE, Andersson MB et al. Endothelin-1 and fibroblast growth factors stimulate the mitogen-activated protein kinase signaling cascade in cardiac myocytes. The potential role of the cascade in the integration of two signaling pathways leading to myocyte hypertrophy. J Biol Chem 1994; 269:1110-1119.
29. Sugden PH. Signalling pathways in cardiac myocyte hypertrophy. Ann Med 2001; 33:611-622.
30. Bueno OF, De Windt LJ, Tymitz KM et al. The MEK1-ERK1/2 signaling pathway promotes compensated hypertrophy in transgenic mice. EMBO J 2000; 19:6341-6350.
31. Chen Z, Gibson TB, Robinson F et al. MAP kinases. Chem Rev 2001; 101:2449-2476.
32. Alvarez E, Northwood IC, Gonzalez FA et al. Pro-Leu-Ser/Thr-Pro is a consensus primary sequence for substrate protein phosphorylation. J Biol Chem 1991; 266:15277-15285.
33. Sharrocks AD. The ETS-domain transcription factor family. Nat Rev Mol Cell Biol 2001; 2:827-837.
34. Frödin M, Gammeltoft S. Role and regulation of 90 kDa ribosomal S6 kinase (RSK) in signal transduction. Mol Cell Endocrin 1999; 151:65-77.
35. Gijon MA, Leslie CC. Regulation of arachidonic acid release and cytosolic phospholipase A_2 activation. J Leukoc Biol 1999; 65:330-336.
36. Chiloeches A, Paterson HF, Marais RM et al. Regulation of Ras.GTP loading and Ras-Raf association in neonatal rat ventricular myocytes by G protein-coupled receptor agonists and phorbol esters. Activation of the ERK cascade by phorbol esters is mediated by Ras. J Biol Chem 1999; 274:19762-19770.
37. Bogoyevitch MA, Marshall CJ, Sugden PH. Hypertrophic agonists stimulate the activities of the protein kinases c-Raf and A-Raf in cultured ventricular myocytes. J Biol Chem 1995; 270:26303-26310.
38. Clerk A, Bogoyevitch MA, Fuller SJ et al. Expression of protein kinase C isoforms during cardiac ventricular development. Am J Physiol 1995; 269:H1087-H1097.
39. Vojtek AB, Der CJ. Increasing complexity of the Ras signaling pathway. J Biol Chem 1998; 273:19925-19928.
40. Takai Y, Sasaki T, Matozaki T. Small GTP-binding proteins. Physiol Rev 2001; 81:153-208.
41. Mason CS, Springer CJ, Cooper RG et al. Serine and tyrosine phosphorylations cooperate in raf-1, but not B-Raf activation. EMBO J 1999; 18:2137-2148.
42. Dhillon AS, Kolch W. Untying the regulation of the Raf-1 kinase. Arch Biochem Biophys 2002; 404:3-9.
43. Fischer TA, Ludwig S, Flory E et al. Activation of cardiac c-Jun NH_2-terminal kinases and p38-mitogen-activated protein kinases with abrupt changes in hemodynamic load. Hypertension 2001; 37:1222-1228.
44. Uozumi H, Hiroi Y, Zou Y et al. gp130 plays a critical role in pressure overload-induced cardiac hypertrophy. J Biol Chem 2001; 276:23115-23119.
45. Esposito G, Prasad SV, Rappacciuolo A et al. Cardiac overexpression of a G_q inhibitor blocks induction of extracellular signal-regulated kinase and c-Jun NH_2-terminal kinase activity in in vivo pressure overload. Circulation 2001; 103:1453-1458.
46. Takeishi Y, Huang Q, Abe J et al. Src and multiple MAP kinase activation in cardiac hypertrophy and congestive heart failure under chronic pressure-overload: Comparison with acute mechanical stretch. J Mol Cell Cardiol 2001; 33:1637-1648.
47. Yamazaki T, Tobe K, Hoh E et al. Mechanical loading activates mitogen-activated protein kinase and S6 peptide kinase in cultured rat cardiac myocytes. J Biol Chem 1993; 268:12069-12076.
48. Cruzalegui FH, Cano E, Treisman R. ERK activation induces phosphorylation of Elk-1 at multiple S/T-P motifs to high stoichiometry. Oncogene 1999; 18:7948-7957.
49. Izumo S, Nadal-Ginard B, Mahdavi V. Protooncogene induction and reprogramming of cardiac gene expression produced by pressure overload. Proc Natl Acad Sci USA 1988; 85:339-343.
50. Aoyagi T, Izumo S. Mapping of the pressure response element of the c-fos gene by direct DNA injection into beating hearts. J Biol Chem 1993; 268:27176-27179.
51. Kyriakis JM, Avruch J. Mammalian mitogen-activated protein kinase signal transduction pathways activated by stress and inflammation. Physiol Rev 2001; 81:807-869.
52. Weston CR, Davis RJ. The JNK signal transduction pathway. Curr Opin Genet Dev 2002; 12:14-21.
53. Gupta S, Barrett T, Whitmarsh AJ et al. Selective interaction of JNK protein kinase isoforms with transcription factors. EMBO J 1996; 15:2760-2770.
54. Sugden PH, Clerk A. "Stress-responsive" mitogen-activated protein kinases (c-Jun N-terminal kinases and p38 mitogen-activated protein kinases) in the myocardium. Circ Res 1998; 83:345-352.
55. Nicol RL, Frey N, Pearson G et al. Activated MEK5 induces serial assembly of sarcomeres and eccentric cardiac hypertrophy. EMBO J 2001; 20:2757-2767.

56. Clerk A, Fuller SJ, Michael A et al. Stimulation of "stress-regulated" mitogen-activated protein kinases (SAPKs/JNKs and p38-MAPKs) in perfused rat hearts by oxidative and other stresses. J Biol Chem 1998; 273:7228-7234.

57. Bogoyevitch MA, Gillespie-Brown J, Ketterman AJ et al. Stimulation of the stress-activated mitogen-activated protein kinases subfamilies in perfused heart. p38/RK mitogen-activated protein kinases and c-jun N-terminal kinases are activated by ischemia/reperfusion. Circ Res 1996; 79:161-172.

58. Komuro I, Kudo S, Yamazaki T et al. Mechanical stretch activates the stress-activated protein kinases in cardiac myocytes. FASEB J 1996; 10:631-636.

59. Yamamoto K, Dang QN, Maeda Y et al. Regulation of cardiomyocyte mechanotransduction by the cardiac cycle. Circulation 2001; 103:1459-1464.

60. Gerdes AM. Remodeling of ventricular myocytes during cardiac hypertrophy and heart failure. J Fla Med Assoc 1992; 79:253-255.

61. Petit V, Thiery JP. Focal adhesions: Structure and dynamics. Biol Cell 2000; 92:477-494.

62. Zamir E, Geiger B. Molecular complexity and dynamics of cell matrix adhesions. J Cell Sci 2001; 114:3583-3590.

63. Sharp WW, Simpson DG, Borg TK et al. Mechanical forces regulate focal adhesion and costamere assembly in cardiac myocytes. Am J Physiol 1997; 273:H546-H556.

64. Leslie NR, Biondi RM, Alessi DR. Phosphoinositide-regulated kinases and phosphoinositide phosphatases. Chem Rev 2001; 101:2365-2380.

65. Buck CA, Horwitz AF. Cell surface receptors for extracellular matrix molecules. Annu Rev Cell Biol 1987; 3:179-205.

66. Akiyama SK. Integrins in cell adhesion and signaling. Hum Cell 1996; 9:181-186.

67. Abram CL, Courtneidge SA. Src family tyrosine kinases and growth factor signaling. Exp Cell Res 2000; 254:1-13.

68. Martin GS. The hunting of the Src. Nat Rev Mol Cell Biol 2001; 2:467-475.

69. Schlaepfer DD, Hauck CR, Sieg DJ. Signaling through focal adhesion kinase. Prog Biophys Mol Biol 1999; 71:435-478.

70. Parsons JT, Martin KH, Slack JK et al. Focal adhesion kinase: A regulator of focal adhesion dynamics and cell movement. Oncogene 2000; 19:5606-5613.

71. Schaller MD. Biochemical signals and biological responses elicited by the focal adhesion kinase. Biochim Biophys Acta 2001; 1540:1-21.

72. Avraham H, Park SY, Schinkmann K et al. RAFTK/Pyk2-mediated cellular signalling. Cell Signal 2000; 12:123-133.

73. Bagrodia S, Cerione RA. PAK to the future. Trends Cell Biol 1999; 9:350-355.

74. Daniels RH, Bokoch GM. p21-activated protein kinase: A crucial component of morphological signaling? Trends Biochem Sci 1999; 24:350-355.

75. Simpson L, Parsons R. PTEN: Life as a tumor suppressor. Exp Cell Res 2001; 264:29-41.

76. O'Neill GM, Fashena SJ, Golemis EA. Integrin signalling: A new Cas(t) enters the stage. Trends Cell Biol 2000; 10:111-119.

77. Turner CE. Paxillin interactions. J Cell Sci 2000; 113:4139-4140.

78. Schaller MD. Paxillin: A focal adhesion-associated adaptor protein. Oncogene 2001; 20:6459-6472.

79. Schlessinger J. Cell signaling by receptor tyrosine kinases. Cell 2000; 103:211-225.

80. Sadoshima J, Xu Y, Slayter HS et al. Autocrine release of angiotensin II mediates stretch-induced hypertrophy of cardiac myocytes in vitro. Cell 1993; 75:977-984.

81. Liang F, Gardner DG. Autocrine/paracrine determinants of strain-activated brain natriuretic peptide gene expression in cultured cardiac myocytes. J Biol Chem 1998; 273:14612-14619.

82. Sugden PH. An overview of endothelin signaling in the cardiac myocyte. J Mol Cell Cardiol 2003, 35:871-876.

83. Sugden PH. Signaling pathways activated by vasoactive peptides in the cardiac myocytes and their role in myocardial pathologies. J Card Fail 2002; 8(Suppl):S359-S369.

84. Liang F, Atakilit A, Gardner DG. Integrin dependence of brain natriuretic peptide promoter activation by mechanical strain. J Biol Chem 2000; 275:20355-20360.

85. Liang F, Kovacic-Milivojevic B, Chen S et al. Signaling mechanisms underlying strain-dependent brain natriuretic peptide gene transcription. Can J Physiol Pharmacol 2001; 79:640-645.

86. Liang F, Lu S, Gardner DG. Endothelin-dependent and -independent components of strain-activated brain natriuretic peptide gene transcription require extracellular signal regulated kinase and p38 mitogen-activated kinase. Hypertension 2000; 35:188-192.

87. Aikawa R, Nagai T, Kudoh S et al. Integrins play a critical role in mechanical stress-induced p38 MAPK activation. Hypertension 2002; 39:233-238.

88. Kuppuswamy D, Kerr C, Narishige T et al. Association of tyrosine-phosphorylated c-Src with the cytoskeleton of hypertrophying myocardium. J Biol Chem 1997; 272:4500-4508.
89. Laser M, Willey CD, Jiang W et al. Integrin activation and focal complex formation in cardiac hypertrophy. J Biol Chem 2000; 275:35624-35630.
90. Shioi T, McMullen JR, Kang PM et al. Akt/protein kinase B promotes organ growth in transgenic mice. Mol Cell Biol 2002; 22:2799-2809.
91. Haq S, Choukroun G, Kang ZB et al. Glycogen synthase kinase-3β is a negative regulator of cardiomyocyte hypertrophy. J Cell Biol 2000; 151:117-130.
92. Hardt SE, Sadoshima J. Glycogen synthase kinase 3β: A novel regulator of cardiac hypertrophy and development. Circ Res 2002; 90:1055-1063.
93. Ramirez MT, Zhao XL, Schulman H et al. The nuclear δB isoform of Ca^{2+}/calmodulin-dependent protein kinase II regulates atrial natriuretic factor gene expression in ventricular myocytes. J Biol Chem 1997; 272:31203-31208.
94. Guan KL, Han M. A G-protein signaling network mediated by an RGS protein. Genes Dev 1999; 13:1763-1767.
95. Booz GW, Day JNE, Baker KM. Interplay between the cardiac renin angiotensin system and JAK/STAT signaling: Role in cardiac hypertrophy. ischemia/reperfusion dysfunction, and heart failure. J Mol Cell Cardiol 2002; 34:1443-1453.
96. Kisseleva T, Bhattacharya S, Braunstein J et al. Signaling through the JAK/STAT pathway, recent advances and future challenges. Gene 2002; 285:1-24.
97. Kaibuchi K, Kuroda S, Amano M. Regulation of the cytoskeleton and cell adhesion by the Rho family GTPases in mammalian cells. Annu Rev Biochem 1999; 68:459-486.
98. Hoshijima M, Sah VP, Wang Y et al. The low molecular weight GTPase Rho regulates myofibril formation and organization in neonatal rat ventricular myocytes. Involvement of Rho kinase. J Biol Chem 1998; 273:7725-7730.
99. Morissette MR, Sah VP, Glembotski CC et al. The Rho effector, PKN, regulates ANF gene transcription in cardiomyocytes through a serum response element. Am J Physiol Heart Circ Physiol 2000; 278:H1769-H1774.
100. Zou Y, Yamazaki T, Nakagawa K et al. Continuous blockade of L-type Ca^{2+} channels suppresses activation of calcineurin and development of cardiac hypertrophy in spontaneously hypertensive rats. Hypertens Res 2002; 25:117-124.
101. Molkentin JD, Lu JR, Antos CL et al. A calcineurin-dependent transcriptional pathway for cardiac hypertrophy. Cell 1998; 93:215-228.
102. Bueno OF, van Rooij E, Molkentin JD et al. Calcineurin and hypertrophic heart disease: Novel insights and remaining questions. Cardiovasc Res 2002; 53:806-821.
103. De Windt LJ, Lim HW, Bueno OF et al. Targeted inhibition of calcineurin attenuates cardiac hypertrophy in vivo. Proc Natl Acad Sci USA 2001; 98:3322-3327.
104. Vanhaesebroeck B, Alessi DR. The PI3K-PDK1 connection: More than just a road to PKB. Biochem J 2000; 346:561-576.
105. Antos CL, McKinsey TA, Frey N et al. Activated glycogen synthase-3β suppresses cardiac hypertrophy in vivo. Proc Natl Acad Sci USA 2002; 99:907-912.
106. Badorff C, Ruetten H, Mueller S et al. Fas receptor signaling inhibits glycogen synthase kinase 3β and induces cardiac hypertrophy following pressure overload. J Clin Invest 2002; 109:373-381.
107. Sheng Z, Knowlton K, Chen J et al. Cardiotrophin 1 (CT-1) inhibition of cardiac myocyte apoptosis via a mitogen-activated protein kinase-dependent pathway. Divergence from downstream CT-1 signals for myocardial cell hypertrophy. J Biol Chem 1997; 272:5783-5791.
108. Pan J, Fukuda K, Kodama H et al. Role of angiotensin II in activation of the JAK/STAT pathway induced by acute pressure overload in the rat heart. Circ Res 1997; 81:611-617.
109. Pan J, Fukuda K, Saito M et al. Mechanical stretch activates the JAK/STAT pathway in rt cardiomyocytes. Circ Res 1999; 84:1127-1136.
110. Suchyna TM, Johnson JH, Hamer K et al. Identification of a peptide toxin from Grammotola spatulata spider venom that blocks cation-selective stretch-activated channels. J Gen Physiol 2000; 115:583-598.

CHAPTER 9

Mechanotransduction of the Endocrine Heart:
Paracrine and Intracellular Regulation of B-Type Natriuretic Peptide Synthesis

Sampsa Pikkarainen, Heikki Tokola and Heikki Ruskoaho*

Abstract

Cardiac overload initiates a process, which aims to maintain and adapt cardiovascular system to altered hemodynamics. In adults, myocardial mass increases mainly due to enlargement of individual myocytes (for reviews, see refs. 1,2). Cardiac pressure overload in conditions such as aortic stenosis or hypertension, results in parallel addition of sarcomeres and increases width of myocytes, which in turn, augment left ventricular wall thickness.[2] However, when mechanical and neurohumoral stress are sustained, the adaptive mechanisms eventually fail and further myocardial remodelling leads to ventricular dilation and impairment of cardiac contractile function. Cardiac output reduces until being inadequate to maintain efficient blood circulation of the whole organism and the syndrome of congestive heart failure occurs.[2,3] At the cellular level, the cardiac growth and failure is due to a complex pattern of signaling mechanisms and molecules. In 1980s, identification of genes associated with cardiac hypertrophy were accompanied by the discovery of natriuretic peptides in the heart.[4,5] Since then, this has been followed by characterization of regulatory mechanisms in natriuretic peptide secretion and synthesis and further insight of the signaling mechanisms and of the development of cardiac hypertrophy has been achieved.

Natriuretic Peptide Family

Separate observations preceded the isolation of atrial natriuretic peptide (ANP), including the identification of granules in atrial myocytes and the finding that elevated left atrial pressure stimulated urine output. Later, it was found that abundance of atrial granules was regulated by age or by changes in dietary sodium and water.[5] The functional role of these granules became elucidated by De Bold et al who injected atrial extract intravenously into intact rats, which produced natriuresis and diuresis.[4] This was followed by identification of ANP and its precursors in atrial tissue.[6-8] After characterization of ANP, structurally and functionally related mammalian peptides, B-type and C-type natriuretic peptides were identified.[9,10] A common structural feature shared by these three natriuretic peptides is the conserved 17 amino acid ring structure.[5] During rodent embryonic and neonatal development, ANP is expressed in atrial and ventricular myocytes, but in adults atria contain 100-150 and 1000 times more ANP mRNA and protein, respectively, than the left ventricle.[5] Similarly, in adult rats BNP mRNA levels are higher in atria than in ventricles, but when related proportioned to the tissue weight,

*Contributing Author: Heikki Ruskoaho—Department of Pharmacology and Toxicology, and Biocenter Oulu, University of Oulu, P.O. Box 5000, 90014 Oulun yliopisto, Oulu, Finland. Email: heikki.ruskoaho@oulu.fi

Cardiac Mechanotransduction, edited by Matti Weckström and Pasi Tavi.
©2007 Landes Bioscience and Springer Science+Business Media.

total amount of mRNA is threefold higher in ventricles.[11] In adult human tissues, BNP mRNA has been detected in the central nervous system, lung, thyroid, adrenal, kidney, spleen, small intestine, ovary, uterus and striated muscle.[12] However, in adult human tissues, the level of extracardiac ANP and BNP mRNAs have been found approximately 10-100 times lower than in ventricle, and in rats, ANP and BNP transcripts have been detected even less in extracardiac rat tissues (1000 times less than in ventricles).[12] Distinct from expression of ANP and BNP in the heart, the production of CNP in rats and humans was reported to be limited to neural system or to a lesser extent, kidney and gastrointestinal tract.[13] However, CNP generation also by cultured endothelial cells has been reported to increase in response to increased laminar flow (shear stress).[14] In addition, family of natriuretic peptides has increased due to identification of related peptides from the salmon and green mamba.[15,16]

All the natriuretic peptides exert their biological functions via binding to specific natriuretic peptide receptors-A and -B (NPR$_A$, NPR$_B$).[17,18] NPR$_{A/B}$ are coupled to guanylyl cyclase and ligand binding to the receptor increases intracellular levels of cGMP.[18] NPR$_{A/B}$ are both present in the adrenal glands and the kidney, but NPR$_A$ receptor is the most abundant type in large blood vessels, while there are also some NPR$_B$ receptors, which predominate in the brain.[19] Due to the affinities of the receptors for different natriuretic peptides, the primary ligands for NPR$_A$ are ANP and BNP, whereas NPR$_B$ binds rather CNP.[5,20] The third receptor identified thus far, natriuretic peptide receptor-C (NPR$_C$), has not well-defined second messengers, and is rather internalized and degraded after ligand binding, suggesting a role as a clearance receptor.[17] Biological effects of natriuretic peptides antagonize those of renin- angiotensin II (Ang II)- aldosterone system: natriuretic peptides decrease blood pressure, increase water and salt excretion and promote vasodilation and inhibit the release or actions of pressor hormones such as Ang II, endothelin-1 (ET-1), renin and vasopressin (for reviews, see refs. 5,19).

Characteristic for the biosynthesis of both ANP and BNP is the production of initial immature peptides, which are then cleaved to release biologically active hormones into circulation.[5] In atrial myocytes, after cleavage of a 25 amino acid signal peptide, the resulting 126 amino acid proANP is stored into granules, and during or soon after exocytosis processed by cardiac serine-protease corin[21,22] into N-terminal fragment of 98 amino acids and to the biologically active ANP of 28 C-terminal amino acids.[5] In contrast, while preproBNP is subjected to analogical stepwise cleavage, the main storage form is the circulating and mature BNP of 45 or 32 amino acids in rats or humans, respectively.[23] In response to acute atrial stretch, the release of ANP and BNP was reported to occur simultaneously,[24] suggesting a similar and regulated secretory pathway for both peptide hormones in mechanically loaded atrial tissue. However, ventricular myocytes contain only few granules[5] and rather a constitutive secretory pathway has been suggested to release the natriuretic peptides from ventricular tissue.[25]

Activation of BNP Synthesis by Mechanical Load

Expression of a gene and the resulting protein, are generally subject to a regulation process at multiple steps including gene transcription, RNA processing and transport, mRNA degradation, stabilization and translation as well as at the level of post-translational processing of the generated protein. In the context of BNP gene, the regulation process results in a tissue-specific expression pattern accompanied by an induction of gene expression in overloaded heart. Mechanical stress alters the phenotype of individual cardiac myocytes profoundly, including hypertrophy and selectively regulated gene expression.[26] Initiation of hypertrophic growth is accompanied by a rapid and transient expression of immediate early genes (e.g., c-*jun*, c-*fos*, c-*myc* and *Egr-1*)[27,28] followed by activation of a pattern of cardiac genes, including ANP and BNP,[5,29] α-skeletal actin (α-skA), β-myosin heavy chain (β-MHC),[30] whereas relative expression of different isoforms of genes is simultaneously decreased (i.e., α-MHC).[3,31]

The rapid inducibility of BNP gene transcription resembles that of immediate early genes, such as c-*fos*.[32] In response to arginine-vasopressine (AVP)- or phenylephrine (PE) infusion BNP mRNA increases already at 1-2 hours in the left atrium and ventricle of spontaneosly hypertensive rats.[32,33] In addition, mechanical loading of perfused rat atrium elevates the levels of BNP mRNA acutely at 1.5-2 hours, an effect that can be blocked by a transcriptional inhibitor, actinomycin D, but not by protein synthesis inhibitor, cycloheximide.[34,35] Similarly, mechanical stretch of cultured cardiomyocytes increases BNP mRNA levels mainly via transcriptional activation.[36] Notably, experiments using primary cell culture show that cardiac myocytes are able to respond to mechanical stretch by increasing BNP secretion and gene expression without neurohumoral regulation.[36] BNP gene expression has been reported to increase in the hypertrophied right ventricle produced by pulmonary artery banding, whereas simultaneously BNP mRNA levels remained unchanged in the left ventricle.[37] Given that BNP gene transcription as well as its release into circulation is activated by cardiac overload, BNP provides an attractive target gene for the study of mechanotransduction and the signaling mechanisms involved.

Local Paracrine and Autocrine Factors and BNP Synthesis during Mechanical Load

Increased load may liberate rapidly local growth promoting factors in the heart. Such autocrine and paracrine factors include angiotensin (Ang II),[38] transforming growth factor-β1 (TGF-β1),[39] basic fibroblast growth factor (bFGF)[40] and endothelin-1 (ET-1).[41] The basic hypothesis underlying this phenomenon is that during increased mechanical load cardiomyocytes and adjacent nonmyocytes (i.e., fibroblasts and endothelial cells) release factors that may bind on cell membrane receptors of cardiomyocytes and activate growth promoting signaling mechanisms and gene expression. Cyclic mechanical stretch of cultured cardiomyocytes has been shown to elevate mRNA levels of angiotensinogen, angiotensin converting enzyme and ET-1.[39] In similar model, mechanical stretch has been reported to increase the release of Ang II within 10-30 minutes[38] and the blockade of angiotensin II type 1a receptor (AT_{1a}) by CV-11974 was shown to inhibit activation of extracellular signal regulated kinase (ERK) by 70%.[42] Losartan, an AT_{1a} blocker, has been shown to inhibit induction human BNP promoter activity in stretched cultured cardiomyocytes.[39] In addition, it has been suggested that in the heart, locally released ET-1 may mediate, at least in part, the hypertrophic responses of myocytes to Ang II or TGF-β1.[39,43,44] In agreement with this hypothesis, pharmacological blockade of endothelin-A (ET_A) receptor with BQ-123 inhibited Ang II and stretch-activated BNP gene transcription by 50 %, while AT_{1a} blockade had no effect on ET-1 induced BNP gene activation. However, endogenous production of ET-1 in cardiomyocytes is relatively low and addition of exogenous ET-1 into culture medium results in potentiation of BNP promoter activity in stretched cultured cardiomyocytes.[39] By using pharmacological blockade of receptors, endogenous ET-1, but not Ang II, has been found to participipate to the regulation of basal and stretch activated BNP transcription in the rat atrium, but not in ventricle.[33] In addition, the transgenic approaches have provided evidence that while Ang II may mediate, it is not requisite for cardiac mechanotransduction, since AT_{1a} null mice develop marked hypertrophy in response to pressure overload[45] and mechanical stretch may activate mitogen activated protein kinases (MAPKs) in cultured angiotensinogen or AT_{1a} deficient cardiomyocytes via alternative mechanisms.[46,47]

Given the activity of mRNA synthesis is reflected by the mRNA levels, it should be noted that the stability of transcript is additionally a critical factor affecting on the mRNA levels and the synthesis of a given peptide. The 3'-untranslated region of BNP mRNA contains several AUUA sequences[48] that may be involved in the translation dependent rapid mRNA degradation.[49] While mechanical load appears to induce BNP levels mainly via transcriptional activation in cultured cardiomyocytes[36] and in stretched rat atrial preparation,[35] α-adrenergic stimulation (PE) of cultured cardiomyocytes involves stabilization of BNP transcripts via protein kinase C (PKC) and ERK.[50] Comparison between BNP gene transcription and mRNA levels has been made during 2-week hemodynamic overload induced by Ang II infusion in rats.[51] This was accomplished by using rat BNP promoter driven reporter plasmid injected into rat myocardium and standard northern blot analysis for BNP mRNA levels.[51] It was demonstrated that transient early induction of BNP promoter activity preceded the increase in BNP mRNA levels. Later the injected BNP promoter was reactivated starting from 3 days and sustaining for 2 weeks, but the BNP mRNA levels remained at the level of control animals.[51] Therefore, induction of BNP gene expression is due to a balance between transcriptional activation and mRNA stability, mechanisms that may either act in parallel or antagonize each others.

Cytosolic Mechanotransduction on BNP Gene

Studies using mechanically stretched myocytes and myocardium undergoing hypertrophic remodelling have demonstrated altered electrophysiological properties, that has led to a hypothesis that the cell membrane mechanosensitive or stretch activated (SA) ion channels participitate in the conversion of physical forces and neurohumoral stimuli to cardiac contractility, gene expression and ultimately to hypertrophic phenotype (for reviews, see refs. 1,52) (Fig. 1). Activation of cell membrane SA channels increases Ca^{2+} influx to myocyte leading to further intracellular Ca^{2+} release and increase in intracellular Ca^{2+}.[53] Studies have been conducted by using trivalent lanthanide gadolinium (Gd^{3+}) that inhibits voltage-gated Ca^{2+}-channels and some of the SA channels, while not all the potential SA channels are sensitive to Gd^{3+}.[54] In isolated superperfused rat atria stretched by increasing intra-atrial pressure, Gd^{3+} has been shown to inhibit increased ANP release and BNP mRNA levels,[55] while in cultured myocytes, mechanical stretch activates protein synthesis and c-fos gene expression in a Gd^{3+}-insensitive manner.[54] In parallel with SA channels that act on the interface of cell membrane and extracellular environment, the integrins may convert physical forces into signaling mechanisms and second messengers in cardiac myocytes. Integrins are noncovalently associated heterodimeric transmembrane receptors composed of α and β subunits that couple components of the extracellular matrix with the actin cytoskeleton (for review, see ref. 56). Integrins are functionally associated with Ras and Rho families of small G proteins as well as with tyrosine kinase signaling mediated by focal adhesion kinase (FAK) and Src tyrosine kinase.[56] Therefore, integrins may transmit mechanical signals to the cytoskeleton and, via coactivated signal transmission, induce BNP gene expression. In support with this, inhibition of tyrosine kinase activity has been reported to suppress the activation of BNP gene in isolated perfused rat atria.[35] Inhibition of integrin and extracellular matrix-interaction by using antibodies raised against $\beta1$, $\beta3$ or $\alpha v\beta5$ integrins has been shown to inhibit ERK, p38 MAPK and, to a lesser extent, c-Jun N-terminal kinase (JNK) pathways and importantly, attenuate induction of human BNP promoter by mechanical strain of cultured cardiomyocytes.[57] In addition, inhibition of FAK may be achieved by overexpressing an inhibitor polypeptide (PTEN) that prevents outside-in signals via integrins-FAK interaction. Overexpression of PTEN in cultured neonatal rat cardiac myocytes has been shown to abolish the activation of p38 MAPK by mechanical stretch.[58] More recently, when pressure overload was induced by aortic banding to transgenic mice with a deletion of muscle-specific integrin $\beta1$-interacting protein melusin, these mice developed lesser increase in heart weight and had a dilated cardiomyopathy rather than concentric hypertrophy seen in wild type mice.[59] In the hearts of transgenic mice, absense of melusin increased activation of glycogen synthase kinase-3β (GSK-3β), but had no effect on activation of ERK and p38 MAPK by pressure overload.[59]

The complex pattern of factors regulate the heart during hemodynamic overload including mechanical, neural and humoral factors acting in concert. The effects of ET-1 and Ang II are mediated via G protein coupled receptors (GPCR). Binding of agonist to GPCR alters the conformation of the receptor and leads to receptor association with heterotrimeric G proteins within the plasma membrane and subsequent activation of G proteins. Both ET and Ang II receptors have been linked to G_q protein mediated activation of phosphoinositide phospholipase-$\beta2$ (PLC-$\beta2$) leading to a generation of inositol-trisphosphate (IP_3) and diacylglycerol (DAG), accompanied by mobilization of calcium and activation of PKC,[60,61] respectively. Dissociated G$\beta\gamma$ has a potential to activate small GTP binding protein Ras and initiate tyrosine kinase cascade leading to activation of MAPK.[62] Inhibition of PKC by chelerythrine blocks the stretch-induced activation of ERK as well as increase in BNP gene expression in cultured ventricular cardiomyocytes and right atrium of isolated perfused rat

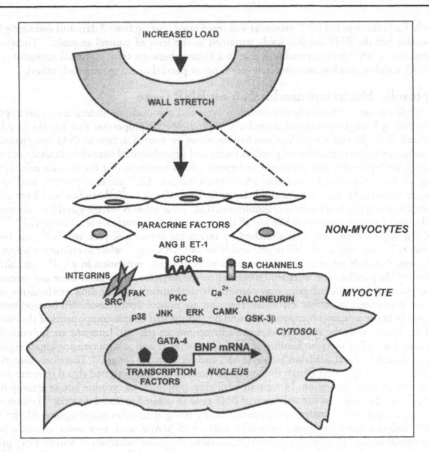

Figure 1. Mechanotransductive pathways regulating BNP gene expression. For abbreviations and symbols, see text.

heart.[35,36] In cultured cardiomyocytes, the overexpression of PKC isozymes has been reported to increase transcription of hypertrophy associated genes, such as ANP and BNP.[63,64] Moreover, expressing dominant interfering forms of Ras and mitogen activated protein kinase kinase-1/2 (MKK1/2), the downstream targets of PKC, suppress stretch-activated BNP transcription suggesting a role for PKC-Ras-MKK1/2-ERK pathway in the regulation of BNP gene expression.[36]

Importantly, all major MAPK cascades beside ERK cascade, i.e., ERK, JNK and p38 MAPK pathways, are activated after the application of either mechanical strain or ET-1 in cardiomyocyte culture.[26,41,64-67] The ERK and p38 MAPK pathways have been shown to be involved in both ET-1-dependent and -independent components of the mechanical strain-induced human BNP promoter with ERK predominating in the ET-1-dependent pathways and p38 MAPK in the ET-1-independent pathways.[67] We have recently demonstrated that activation of rat BNP promoter by ET-1 is inhibited by 60% with overexpression of dominant interfering mutant of MKK6 (an upsteam kinase of p38 MAPK), but not by dominant negative mutants of MKK1 or JNK1.[68] Interestingly, we detected over 4-fold induction of rat BNP promoter in response to L-type Ca^{2+}-channel agonist, and this response was equally inhibited by inhibition of all three MAPK cascades.[68] Furthermore, inhibition of intracellular Ca^{2+}-binding protein calmodulin (CaM) attenuates strain-activated BNP transcription as does the inhibition of Ca^{2+}/CaM kinase (CaMK) although to lesser extent in cultured cardiomyocytes.[36] This observation

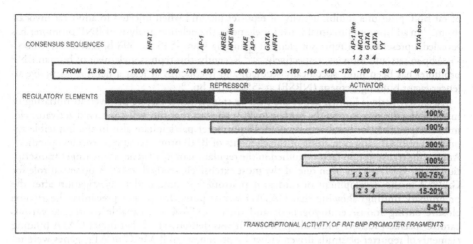

Figure 2. Regulatory elements rat BNP promoter and their impact on basal activity (%). For more detailed explanation, see text.

raises the possibility that Ca^{2+}-CaM acts also through other downstream targets such as calcineurin. Calcineurin has been reported to be activated during load-induced cardiac hypertrophy, communicate with PKC and JNK pathways, and suggested to link intracellular Ca^{2+}-signaling pathways to hypertrophic response.[69] At the low level of overexpression we were able to detect activation of intact BNP promoter by only p38 MAPK, but not with ERK or JNK cascades.[68] However, beside ERK and p38 MAPK, also overexpression of JNK has been shown to potentiate human BNP promoter activity in cultured cardiomyocytes.[36] Vice versa, in the same model the overexpression of dominant negative JNK inhibited mechanical stretch induced BNP transcription.[36] Our findings suggest that p38 MAPK may mediate induction of rat BNP transcription in vitro to ET-1, but in general, induction of BNP gene by more robust stimuli, such as mechanical load and increased intracellular Ca^{2+} is not restricted to p38 MAPK mediated mechanism.[68] Thus it appears that all three MAPK cascades activated in parallel with Ca^{2+} regulated signaling pathways may orchestrate the activation of BNP gene transcription during hemodynamic load.

Nuclear Mechanotransduction on BNP Gene

Activity of BNP gene transcription in cardiac myocytes is regulated by a complex array of intracellular signaling cascades, which are switched on and off by numerous extracellular stimuli. However, in the context of a gene transcription, a wide pattern of signaling mechanisms converges to a limited number of regulatory elements of the gene. Thus, nuclear transcription factors provide a point of convergence in the signaling network and may prove attractive targets for drug research.[70]

5′-flanking regulatory region (promoter) of BNP gene contains a number of potential consensus sequences for binding of transcription factors (Fig. 2). The deletion constructs of BNP promoter together with transfection of these promoter-driven reporter genes (luciferase) into cardiac myocytes has led to an identification of crucial elements in BNP gene regulation.[63,71-73] BNP promoter contains a proximal activator element that is composed of an activator protein-1 (AP-1) like element, a M-CAT and two GATA elements in rats and of two M-CAT elements and a GATA element in humans.[71-73] This proximal activator element of BNP is required for basal promoter activity and is also sufficient to direct activity of the promoter fragment to cardiac myocytes.[71,73] More specifically, the mutations of M-CAT and GATA elements have been shown to inhibit baseline BNP promoter activity in cultured neonatal rat cardiac myocytes.[71,73,74] Similarly, mutations and deletions of AP-1-like and GATA elements of rat BNP promoter inhibit activity of reporter plasmid when injected to adult rat myocardium.[75] In addition to proximal activator element, the deletion analysis of BNP promoter has revealed a presence of a repressor element spanning from -535 to -398 bp and from -906 to -500 bp in rats and humans, respectively.[72,73] Recently, this repressor element of human BNP promoter has been more specifically studied, and it contains at least a neuron restrictive silencer factor binding element (NRSE) at -552- -522 bp.[76]

Relatively little is known about the nuclear transcription factors involved in initiating or maintaining the response to the cardiac load. It appears that the well-conserved activator elements responsible for basal activity of BNP promoter participitate also in the inducible expression of BNP. However, some of the elements of BNP promoter appear species-specific.[29] Among the transcription factors studied in the regulation of BNP gene, a zinc finger transcription factor GATA-4 has been one of the most extensively studied so far. A potential role for GATA-4 in the development of cardiac hypertrophy became under investigation after the preliminary findings showing that GATA-4 was required for proper precardiac heart tube closure during mouse embryogenesis and since GATA-4 was capable to induce various cardiac-specific promoters including BNP.[69] It was demonstrated that intact GATA binding elements of reporter plasmids driven either by promoters of β-MHC or AT_{1a} genes were required for the full induction in pressure overloaded rat hearts.[77,78] Similar method has been utilized to identify the mechanisms that couple hemodynamic stress to alterations in BNP gene expression. Injected DNA constructs containing the rat BNP promoter linked to a reporter gene were injected into the ventricular myocardium of rats, which underwent bilateral nephrectomy for a one-day period or were sham-operated.[75] In this model, ventricular BNP gene expression was induced about 4-fold at the mRNA levels and the GATA elements at -90 bp were both necessary and sufficient to confer transcriptional activation of BNP gene in response to hemodynamic stress.[75] In the heart, signaling cascades responsible for the activation of GATA-4 during cardiac overload may involve locally released paracrine mediators. Hautala et. al. have shown that GATA-4 is activated at a very early stage of pressure overload, well before the development of left ventricular hypertrophy.[79] Activation of GATA-4 during pressure overload was blocked with mixed ET receptor blocker, bosentan, suggesting a role for endogenously released ET-1.[79] In perfused rat hearts, mechanical loading has been reported to activate GATA-4 binding through endogenous ET-1 and Ang II, and when infused, both substances were sufficient to induce GATA-4 DNA binding activity.[80] In static cardiomyocyte culture, ET-1 and PE are also sufficient to increase GATA-4 DNA binding activity.[79,81,82] Moreover, we and others have recently shown that activation of GATA-4 by stimulation with hypertrophic agonists is mediated via MAPK regulated serine-phosphorylation of GATA-4.[81,83,84] In static cardiomyocyte culture, M-CAT and duplicate GATA elements have been reported to mediate the PE-induced activity to a similar extent as they mediate basal rat BNP promoter activity (98% and 60%, respectively).[63] However, we have found that in the absence of mechanical stretch, blockade of GATA-4 DNA binding activity is not sufficient to prevent activation of BNP transcription by ET-1,[82] suggesting that, at least in cultured cardiomyocytes, ET-1 may recruit also alternative signaling mechanisms. This was recently studied more detail, and in fact, ET-1 was shown regulate rat BNP transcription exclusively via p38 MAPK mediated activation of ETS-like gene-1 (Elk-1) transcription factor.[68] In general, multiple line of evidence suggest that GATA-4 may have a role in cardiac adaptation to increased load, and the overexpression of GATA-4 or GATA-6 has been shown to activate hypertrophic growth in cultured cardiomyocytes and in the hearts of transgenic mice.[85] In addition to growth promoting effects, GATA-4 may protect cardiac myocytes from apoptosis.[86]

GATA-4 has been shown to cooperate with a number of other transcription factors in vitro.[69] Specifically, C-terminal Zn-finger domain of GATA-4 that is required for DNA binding has been found to interact physically and to be required for transcriptional coactivation in concert with several transcription factors, including GATA-6, dHAND, MEF-2, nuclear factor of activated T-cells 3 (NFAT3), Nkx-2,5, p300, SRF and YY1.[87-92] In contrast, N-terminal Zn-finger domain of GATA-4 was found to associate with FOG-2 transcription factor resulting in repression of GATA-4 mediated ANP and BNP transcription.[93,94] However, altogether the hierarchy and regulation of GATA-4 interactions is relatively poorly characterized, and only a little is known of the significance of these in vitro interactions during cardiac hypertrophy. We have provided first evidence for such interaction by using rat BNP promoter with site-specific mutations of GATA and Nkx-2,5 binding element (NKE) in stretched culture of cardiomyocytes. Mutation of GATA binding sequence of BNP inhibited stretch-response by 40% and was almost completely abolished when GATA mutation was combined with NKE site mutation.

To directly assess the role GATA-4 in hypertrophic response of stretched myocytes, we have used adenoviral antisense strategy to attenuate GATA-4 protein levels.[93] Previously, adenoviral GATA-4-as has been shown to downregulate cardiac gene expression, including BNP, α- and β-MHC and cardiac Troponin I.[93] The lack of GATA-4 did not inhibit all cardiac genes, since cardiac α-actin and MLC-1 mRNAs were not decreased.[93] We have found that decreased GATA-4 protein levels inhibit stretch-induced BNP mRNA levels and sarcomeric protein assembly. Importantly, the attenuation of protein levels of GATA-4 may inhibit also the potential interactions and thus transactivation with cofactors of GATA-4. In stretched cultured cardiomyocytes, mRNA levels of GATA-4 increased transiently at 4 hours, while the DNA binding activity was transiently upregulated at 1 hour together with nuclear accumulation of GATA-4 detected already at 15 minutes. Therefore, it is likely that multiple levels of regulation of and via GATA-4 exists. In addition to the MAPK cascades, other upstream regulators of GATA-4 have been identified. GSK-3β-has been shown to act as a negative regulator of cardiac hypertrophy, and its decreased activity has been associated with cardiac hypertrophy (for review, see ref. 95). Interestingly, in a manner analogical to NFAT3,[95] GSK-3β has been shown to phosphorylate and activate nuclear export of GATA-4 in cardiac myocytes leading to inhibition of GATA-4 dependent transcription.[96] Therefore, mechanotransduction via GATA-4 involves a complex pattern of regulation that likely takes place at the level of synthesis, localization and post-transciptional modifications of GATA-4 protein, which in turn, may alter its interactions with a pattern of cofactors.

References

1. Swynghedauw B. Molecular mechanisms of myocardial remodeling. Physiol Rev 1999; 79:215-262.
2. Lorell BH, Carabello BA. Left ventricular hypertrophy: Pathogenesis, detection, and prognosis. Circulation 2000; 102:470-479.
3. Hunter JJ, Chien KR. Signaling pathways for cardiac hypertrophy and failure. N Engl J Med 1999; 341:1276-1283.
4. de Bold AJ, Borenstein HB, Veress AT et al. A rapid and potent natriuretic response to intravenous injection of atrial myocardial extract in rats. Life Sci 1981; 28:89-94.
5. Ruskoaho H. Atrial natriuretic peptide: Synthesis, release, and metabolism. Pharmacol Rev 1992; 44:479-602.
6. Nakayama K, Ohkubo H, Hirose T et al. mRNA sequence for human cardiodilatin-atrial natriuretic factor precursor and regulation of precursor mRNA in rat atria. Nature 1984; 310:699-701.
7. Greenberg BD, Bencen GH, Seilhamer JJ et al. Nucleotide sequence of the gene encoding human atrial natriuretic factor precursor. Nature 1984; 312:656-658.
8. Yamanaka M, Greenberg B, Johnson L et al. Cloning and sequence analysis of the cDNA for the rat atrial natriuretic factor precursor. Nature 1984; 309:719-722.
9. Sudoh T, Kangawa K, Minamino N et al. A new natriuretic peptide in porcine brain. Nature 1988; 332:78-81.
10. Sudoh T, Minamino N, Kangawa K et al. C-type natriuretic peptide (CNP): A new member of natriuretic peptide family identified in porcine brain. Biochem Biophys Res Commun 1990; 168:863-870.
11. Ogawa Y, Nakao K, Mukoyama M et al. Natriuretic peptides as cardiac hormones in normotensive and spontaneously hypertensive rats. The ventricle is a major site of synthesis and secretion of brain natriuretic peptide. Circ Res 1991; 69:491-500.
12. Gerbes AL, Dagnino L, Nguyen T et al. Transcription of brain natriuretic peptide and atrial natriuretic peptide genes in human tissues. J Clin Endocrinol Metab 1994; 78:1307-1311.
13. Komatsu Y, Nakao K, Suga S et al. C-type natriuretic peptide (CNP) in rats and humans. Endocrinology 1991; 129:1104-1106.

14. Chun TH, Itoh H, Ogawa Y et al. Shear stress augments expression of C-type natriuretic peptide and adrenomedullin. Hypertension 1997; 29:1296-1302.
15. Schweitz H, Vigne P, Moinier D et al. A new member of the natriuretic peptide family is present in the venom of the green mamba (Dendroaspis angusticeps). J Biol Chem 1992; 267:13928-13932.
16. Majalahti-Palviainen T, Hirvinen M, Tervonen V et al. Gene structure of a new cardiac peptide hormone: A model for heart- specific gene expression. Endocrinology 2000; 141:731-740.
17. Nakao K, Ogawa Y, Suga S et al. Molecular biology and biochemistry of the natriuretic peptide system. II: Natriuretic peptide receptors. J Hypertens 1992b; 10:1111-1114.
18. Potter LR, Hunter T. Guanylyl cyclase-linked natriuretic peptide receptors: Structure and regulation. J Biol Chem 2001; 276:6057-6060.
19. Levin ER, Gardner DG, Samson WK. Natriuretic peptides. N Engl J Med 1998; 339:321-328.
20. Yandle TG. Biochemistry of natriuretic peptides. J Intern Med 1994; 235:561-576.
21. Yan W, Wu F, Morser J et al. Corin, a transmembrane cardiac serine protease, acts as a pro-atrial natriuretic peptide-converting enzyme. Proc Natl Acad Sci USA 2000; 97:8525-8529.
22. Wu F, Yan W, Pan J et al. Processing of pro-atrial natriuretic peptide by corin in cardiac myocytes. J Biol Chem 2002; 277:16900-16905.
23. Nakao K, Ogawa Y, Suga S et al. Molecular biology and biochemistry of the natriuretic peptide system. I: Natriuretic peptides. J Hypertens 1992a; 10:907-912.
24. Mäntymaa P, Vuolteenaho O, Marttila M et al. Atrial stretch induces rapid increase in brain natriuretic peptide but not in atrial natriuretic peptide gene expression in vitro. Endocrinology 1993; 133:1470-1473.
25. Wei CM, Heublein DM, Perrella MA et al. Natriuretic peptide system in human heart failure. Circulation 1993; 88:1004-1009.
26. Sadoshima J, Izumo S. Mechanical stretch rapidly activates multiple signal transduction pathways in cardiac myocytes: Potential involvement of an autocrine/paracrine mechanism. EMBO J 1993; 12:1681-1692.
27. Komuro I, Kaida T, Shibazaki Y et al. Stretching cardiac myocytes stimulates protooncogene expression. J Biol Chem 1990; 265:3595-3598.
28. Yamazaki T, Komuro I, Yazaki Y. Molecular mechanism of cardiac cellular hypertrophy by mechanical stress. J Mol Cell Cardiol 1995c; 27:133-140.
29. Tokola H, Hautala N, Marttila M et al. Mechanical load-induced alterations in B-type natriuretic peptide gene expression. Can J Physiol Pharmacol 2001; 79:646-653.
30. Izumo S, Lompre AM, Matsuoka R et al. Myosin heavy chain messenger RNA and protein isoform transitions during cardiac hypertrophy. Interaction between hemodynamic and thyroid hormone-induced signals. J Clin Invest 1987; 79:970-977.
31. Nakao K, Minobe W, Roden R et al. Myosin heavy chain gene expression in human heart failure. J Clin Invest 1997; 100:2362-2370.
32. Magga J, Marttila M, Mäntymaa P et al. Brain natriuretic peptide in plasma, atria, and ventricles of vasopre. Endocrinology 1994; 134:2505-2515.
33. Magga J, Vuolteenaho O, Marttila M et al. Endothelin-1 is involved in stretch-induced early activation of B-type natriuretic peptide gene expression in atrial but not in ventricular myocytes: Acute effects of mixed ET(A)/ET(B) and AT1 receptor antagonists in vivo and in vitro. Circulation 1997a; 96:3053-3062.
34. Bruneau BG, Piazza LA, de Bold AJ. BNP gene expression is specifically modulated by stretch and ET-1 in a new model of isolated rat atria. Am J Physiol 1997; 273:H2678-H2686.
35. Magga J, Vuolteenaho O, Tokola H et al. Involvement of transcriptional and posttranscriptional mechanisms in cardiac overload-induced increase of B-type natriuretic peptide gene expression. Circ Res 1997b; 81:694-702.
36. Liang F, Wu J, Garami M et al. Mechanical strain increases expression of the brain natriuretic peptide gene in rat cardiac myocytes. J Biol Chem 1997; 272:28050-28056.
37. Adachi S, Ito H, Ohta Y et al. Distribution of mRNAs for natriuretic peptides in RV hypertrophy after pulmonary arterial banding. Am J Physiol 1995; 268:H162-H169.
38. Sadoshima J, Xu Y, Slayter HS et al. Autocrine release of angiotensin II mediates stretch-induced hypertrophy of cardiac myocytes in vitro. Cell 1993; 75:977-984.
39. Liang F, Gardner DG. Autocrine/paracrine determinants of strain-activated brain natriuretic peptide gene expression in cultured cardiac myocytes. J Biol Chem 1998; 273:14612-14619.
40. Kaye D, Pimental D, Prasad S et al. Role of transiently altered sarcolemmal membrane permeability and basic fibroblast growth factor release in the hypertrophic response of adult rat ventricular myocytes to increased mechanical activity in vitro. J Clin Invest 1996; 97:281-291.
41. Yamazaki T, Komuro I, Kudoh S et al. Endothelin-1 is involved in mechanical stress-induced cardiomyocyte hypertrophy. J Biol Chem 1996; 271:3221-3228.

42. Yamazaki T, Komuro I, Kudoh S et al. Angiotensin II partly mediates mechanical stress-induced cardiac hypertrophy. Circ Res 1995a; 77:258-265.
43. Ito H, Hirata Y, Adachi S et al. Endothelin-1 is an autocrine/paracrine factor in the mechanism of angiotensin II-induced hypertrophy in cultured rat cardiomyocytes. J Clin Invest 1993; 92:398-403.
44. Harada M, Itoh H, Nakagawa O et al. Significance of ventricular myocytes and nonmyocytes interaction during cardiocyte hypertrophy: Evidence for endothelin-1 as a paracrine hypertrophic factor from cardiac nonmyocytes. Circulation 1997; 96:3737-3744.
45. Harada K, Komuro I, Shiojima I et al. Pressure overload induces cardiac hypertrophy in angiotensin II type 1A receptor knockout mice. Circulation 1998; 97:1952-1959.
46. Nyui N, Tamura K, Mizuno K et al. Stretch-induced MAP kinase activation in cardiomyocytes of angiotensinogen-deficient mice. Biochem Biophys Res Commun 1997; 235:36-41.
47. Kudoh S, Komuro I, Hiroi Y et al. Mechanical stretch induces hypertrophic responses in cardiac myocytes of angiotensin II type 1a receptor knockout mice. J Biol Chem 1998; 273:24037-24043.
48. Porter JG, Arfsten A, Palisi T et al. Cloning of a cDNA encoding porcine brain natriuretic peptide. J Biol Chem 1989; 264:6689-6692.
49. Shaw G, Kamen R. A conserved AU sequence from the 3' untranslated region of GM-CSF mRNA mediates selective mRNA degradation. Cell 1986; 46:659-667.
50. Hanford DS, Glembotski CC. Stabilization of the B-type natriuretic peptide mRNA in cardiac myocytes by alpha-adrenergic receptor activation: Potential roles for protein kinase C and mitogen-activated protein kinase. Mol Endocrinol 1996; 10:1719-1727.
51. Suo M, Hautala N, Földes G et al. Posttranscriptional control of BNP gene expression in angiotensin II- induced hypertension. Hypertension 2002; 39:803-808.
52. Tavi P, Laine M, Weckström M et al. Cardiac mechanotransduction: From sensing to disease and treatment. Trends Pharmacol Sci 2001; 22:254-260.
53. Sigurdson W, Ruknudin A, Sachs F. Calcium imaging of mechanically induced fluxes in tissue-cultured chick heart: Role of stretch-activated ion channels. Am J Physiol 1992; 262:H1110-H1115.
54. Sadoshima J, Izumo S. The cellular and molecular response of cardiac myocytes to mechanical stress. Annu Rev Physiol 1997; 59:551-571.
55. Laine M, Id L, Vuolteenaho O et al. Role of calcium in stretch-induced release and mRNA synthesis of natriuretic peptides in isolated rat atrium. Pflugers Arch 1996; 432:953-960.
56. Parsons JT. Integrin-mediated signalling: Regulation by protein tyrosine kinases and small GTP-binding proteins. Curr Opin Cell Biol 1996; 8:146-152.
57. Liang F, Atakilit A, Gardner DG. Integrin dependence of brain natriuretic peptide gene promoter activation by mechanical strain. J Biol Chem 2000a; 275:20355-20360.
58. Aikawa R, Nagai T, Kudoh S et al. Integrins play a critical role in mechanical stress-induced p38 MAPK activation. Hypertension 2002; 39:233-238.
59. Brancaccio M, Fratta L, Notte A et al. Melusin, a muscle-specific integrin beta1-interacting protein, is required to prevent cardiac failure in response to chronic pressure overload. Nat Med 2003; 9:68-75.
60. Nishizuka Y. Turnover of inositol phospholipids and signal transduction. Science 1984; 225:1365-1370.
61. Nishizuka Y. Studies and perspectives of protein kinase C. Science 1986; 233:305-312.
62. van Biesen T, Hawes BE, Luttrell DK et al. Receptor-tyrosine-kinase- and G beta gamma-mediated MAP kinase activation by a common signalling pathway. Nature 1995; 376:781-784.
63. Thuerauf DJ, Glembotski CC. Differential effects of protein kinase C, Ras, and Raf-1 kinase on the induction of the cardiac B-type natriuretic peptide gene through a critical promoter-proximal M-CAT element. J Biol Chem 1997; 272:7464-7472.
64. Sugden PH, Clerk A. Cellular mechanisms of cardiac hypertrophy. J Mol Med 1998; 76:725-746.
65. Yamazaki T, Komuro I, Kudoh S et al. Mechanical stress activates protein kinase cascade of phosphorylation in neonatal rat cardiac myocytes. J Clin Invest 1995b; 96:438-446.
66. Liang F, Gardner DG. Mechanical strain activates BNP gene transcription through a p38/NF-kappaB-dependent mechanism. J Clin Invest 1999; 104:1603-1612.
67. Liang F, Lu S, Gardner DG. Endothelin-dependent and -independent components of strain-activated brain natriuretic peptide gene transcription require extracellular signal regulated kinase and p38 mitogen-activated protein kinase. Hypertension 2000b; 35:188-192.
68. Pikkarainen S, Tokola H, Kerkelä R et al. Endothelin-1 specific activation of B-type natriuretic peptide gene via p38 mitogen activated protein kinase and nuclear ETS factors. J Biol Chem 2002b.
69. Molkentin JD. The zinc finger-containing transcription factors GATA-4, -5, and -6. Ubiquitously expressed regulators of tissue-specific gene expression. J Biol Chem 2000a; 275:38949-38952.
70. Emery JG, Ohlstein EH, Jaye M. Therapeutic modulation of transcription factor activity. Trends Pharmacol Sci 2001; 22:233-240.

71. Grepin C, Dagnino L, Robitaille L et al. A hormone-encoding gene identifies a pathway for cardiac but not skeletal muscle gene transcription. Mol Cell Biol 1994; 14:3115-3129.
72. Thuerauf DJ, Hanford DS, Glembotski CC. Regulation of rat brain natriuretic peptide transcription. A potential role for GATA-related transcription factors in myocardial cell gene expression. J Biol Chem 1994; 269:17772-17775.
73. Lapointe MC, Wu G, Garami M et al. Tissue-specific expression of the human brain natriuretic peptide gene in cardiac myocytes. Hypertension 1996; 27:715-722.
74. He Q, Mendez M, Lapointe MC. Regulation of the human brain natriuretic peptide gene by GATA-4. Am J Physiol Endocrinol Metab 2002; 283:E50-E57.
75. Marttila M, Hautala N, Paradis P et al. GATA4 mediates activation of the B-type natriuretic peptide gene expression in response to hemodynamic stress. Endocrinology 2001; 142:4693-4700.
76. Ogawa E, Saito Y, Kuwahara K et al. Fibronectin signaling stimulates BNP gene transcription by inhibiting neuron-restrictive silencer element-dependent repression. Cardiovasc Res 2002; 53:451-459.
77. Herzig TC, Jobe SM, Aoki H et al. Angiotensin II type1a receptor gene expression in the heart: AP-1 and GATA-4 participate in the response to pressure overload. Proc Natl Acad Sci USA 1997; 94:7543-7548.
78. Hasegawa K, Lee SJ, Jobe SM et al. cis-Acting sequences that mediate induction of beta-myosin heavy chain gene expression during left ventricular hypertrophy due to aortic constriction [see comments]. Circulation 1997; 96:3943-3953.
79. Hautala N, Tokola H, Luodonpää M et al. Pressure overload increases GATA4 binding activity via endothelin-1. Circulation 2001; 103:730-735.
80. Hautala N, Tenhunen O, Szokodi I et al. Direct left ventricular wall stretch activates GATA4 binding in perfused rat heart: Involvement of autocrine/paracrine pathways. Pflugers Arch 2002; 443:362-369.
81. Liang Q, Wiese RJ, Bueno OF et al. The transcription factor GATA4 is activated by extracellular signal- Regulated kinase 1- and 2-mediated phosphorylation of serine 105 in cardiomyocytes. Mol Cell Biol 2001b; 21:7460-7469.
82. Pikkarainen S, Kerkelä R, Pöntinen J et al. Decoy oligonucleotide characterization of GATA-4 transcription factor in hypertrophic agonist induced responses of cardiac myocytes. J Mol Med 2002a; 80:51-60.
83. Charron F, Tsimiklis G, Arcand M et al. Tissue-specific GATA factors are transcriptional effectors of the small GTPase RhoA. Genes Dev 2001; 15:2702-2719.
84. Kerkelä R, Pikkarainen S, Majalahti-Palviainen T et al. Distinct roles of mitogen activated protein kinase pathways in GATA-4 transcription factor mediated regulation of B-type natriuretic peptide gene. J Biol Chem 2002; 277:13752-13760.
85. Liang Q, De Windt LJ, Witt SA et al. The transcription factors GATA4 and GATA6 regulate cardiomyocyte hypertrophy in vitro and in vivo. J Biol Chem 2001a; 276:30245-30253.
86. Kim Y, Ma AG, Kitta K et al. Anthracycline-induced suppression of GATA-4 transcription factor: Implication in the regulation of cardiac myocyte apoptosis. Mol Pharmacol 2003; 63:368-377.
87. Durocher D, Charron F, Warren R et al. The cardiac transcription factors Nkx2-5 and GATA-4 are mutual cofactors. EMBO J 1997; 16:5687-5696.
88. Molkentin JD, Lu JR, Antos CL et al. A calcineurin-dependent transcriptional pathway for cardiac hypertrophy. Cell 1998; 93:215-228.
89. Belaguli NS, Sepulveda JL, Nigam V et al. Cardiac tissue enriched factors serum response factor and GATA-4 are mutual coregulators. Mol Cell Biol 2000; 20:7550-7558.
90. Morin S, Charron F, Robitaille L et al. GATA-dependent recruitment of MEF2 proteins to target promoters. EMBO J 2000; 19:2046-2055.
91. Bhalla SS, Robitaille L, Nemer M. Cooperative activation by GATA-4 and YY1 of the cardiac B-type natriuretic peptide promoter. J Biol Chem 2001; 276:11439-11445.
92. Dai YS, Cserjesi P, Markham BE et al. The transcription factors GATA4 and dHAND physically interact to synergistically activate cardiac gene expression through a p300- Dependent mechanism. J Biol Chem 2002; 277:24390-24398.
93. Charron F, Paradis P, Bronchain O et al. Cooperative interaction between GATA-4 and GATA-6 regulates myocardial gene expression. Mol Cell Biol 1999; 19:4355-4365.
94. Tevosian SG, Deconinck AE, Cantor AB et al. FOG-2: A novel GATA-family cofactor related to multitype zinc-finger proteins Friend of GATA-1 and U-shaped. Proc Natl Acad Sci USA 1999; 96:950-955.
95. Hardt SE, Sadoshima J. Glycogen synthase kinase-3beta: A novel regulator of cardiac hypertrophy and development. Circ Res 2002; 90:1055-1063.
96. Morisco C, Seta K, Hardt SE et al. Glycogen synthase kinase 3beta regulates GATA4 in cardiac myocytes. J Biol Chem 2001; 276:28586-28597.

Index

A

α-actinin 11, 29, 40, 41, 82, 84, 85
α-adrenergic receptor 95
α1-adrenoceptor 107-109, 111, 113, 116
$α1_A$-adrenoceptor 107-111
$α1_B$-adrenoceptor 107-109, 111
α2-adrenoceptor 107, 108, 111, 112
ACE inhibitor 66, 67, 95
Actin 10, 11, 28, 29, 32, 33, 35, 38, 40-42,
    79, 81, 82, 84, 89, 94, 96, 97, 99, 107,
    127, 135, 137, 141
Actin microfilament 29, 38, 41
Actin-myosin complex 79
Action potential 3, 4, 8, 19-21, 38, 49-56,
    59-64, 73-75, 77, 85, 121
Action potential duration 8, 19, 38, 49, 51,
    53, 54, 56, 64
ADH 4
Adrenergic receptor (AR) 65, 95, 96, 99,
    107-109
Afterdepolarisation 50-52, 60, 62-64
Akt 110, 122, 128, 129
Angiotensin 18, 21, 22, 36, 38, 57, 66, 67,
    81, 94, 95, 106, 115, 116, 128, 135, 136
aPKC 122
Arrhythmia 3, 20, 52, 53, 56, 57, 60-69, 93
AT-1 receptor 18, 21, 95, 128
Athlete's heart 5
Atrial natriuretic factor (ANF) 80, 107, 121
Atrial natriuretic peptide (ANP) 3, 94, 96, 98,
    99, 121, 134, 135, 137, 138, 141
Automaticity 8, 20

B

β-adrenoceptor 65, 107, 108, 111-115
B-type (brain) natriuretic peptide (BNP) 3, 8,
    85, 86, 96, 98, 99, 121, 126, 128,
    134-141

C

c-fos 94-96, 99, 123, 126, 135, 137
c-Jun N-terminal kinase (JNK) 81, 94, 95,
    101, 111, 120, 122, 123, 126, 127,
    137-139
c-myc 94, 135
C-type natriuretic peptide 134
Ca^{2+} transient 3, 20, 21, 38, 94, 121
Calcineurin 94, 98, 99, 101, 122, 128, 139
Calcium overload 62
Calcium transient 53, 54, 79
Calmodulin (CaM) 9, 81, 94, 98, 122, 128,
    129, 138, 139
CaMKII 98
Cardiac dilation 78, 83
Cardiac muscle 1, 2, 28-34, 38, 42, 48, 81,
    85, 97
Cardiomyopathy 35, 36, 39, 61, 63, 65, 66,
    78, 79, 82, 83, 85-87, 93, 99, 137
Charybdotoxin 15, 56
Collagen 10, 21, 35, 39, 57, 80, 81, 93, 120
Compensated hypertrophy 78
Connexin-43 80
Continuum model 38
Contractile apparatus 4, 5, 11
Contraction force 1-3, 121
Control system 3, 57-60
Costamere 11, 29, 40, 41, 81, 82, 85
cPKC 122-124
Crossbridge 28, 30, 32, 33, 42
Cyclic AMP (cAMP) 36, 38, 55-57, 81, 96,
    112-115
Cytokine 93, 94, 96, 99-101, 106, 115, 116,
    129
Cytoskeleton 9-13, 16, 18, 28, 29, 38-42,
    54-57, 59, 61, 66, 68, 78, 79, 81-86, 89,
    97, 121, 127, 137